Automobile Electronics

Japanese Technology Reviews

Editor in Chief

Toshiaki Ikoma, *University of Tokyo, Japan*

Section Editors

Manufacturing Engineering	Fumio Harashima, *University of Tokyo*
Biotechnology	Isao Karube, *University of Tokyo*
Electronics	Toshiaki Ikoma, *University of Tokyo*
New Materials	Hiroaki Yanagida, *University of Tokyo*
Computers and Communications	Yasuo Kato, *NEC Corporation, Kawasaki*
	Kazumoto Iinuma, *NEC Corporation, Kawasaki*
	Tadao Saito, *University of Tokyo*

Manufacturing Engineering

Automobile Electronics
Shoichi Washino

Steel Manufacturing I: Manufacturing System
Tadao Kawaguchi and Kenji Sugiyama

Steel Manufacturing II: Control System
Tadao Kawaguchi and Takatsugu Ueyama

Biotechnology

Production of Nucleotides and Nucleosides by Fermentation
Sadao Teshiba and Akira Furuya

Recent Progress in Microbial Production of Amino Acids
Hitoshi Enei, Kenzai Yokozeki and Kunihiko Akashi

Electronics

MMIC—Monolithic Microwave Integrated Circuits
Yasuo Mitsui

Bulk Crystal Growth Technology
Shin-ichi Akai, Keiichiro Fujita, Masamichi Yokogawa, Mikio Morioka and Kazuhisa Matsumoto

Semiconductor Heterostructure Devices (Electronics)
Masayuki Abe and Naoki Yokohama

This book is part of a series. The publisher will accept continuation orders which may be cancelled at any time and which provide for automatic billing and shipping of each title in the series upon publication. Please write for details.

Automobile Electronics

by
Shoichi Washino
Mitsubishi Electric Corporation
Hyogo, Japan

Gordon and Breach Science Publishers

New York • London • Paris • Montreux • Tokyo • Melbourne

Gordon and Breach Science Publishers

Pos Office Box 786
Cooper Station
New York, New York 10276
United States of America

Post Office Box 197
London WC2E 9PX
England

58, rue Lhomond
75005 Paris
France

Post Office Box 161
1820 Montreux 2
Switzerland

3-14-9, Okubo
Shinjuku-ku, Tokyo
Japan

Private Bag 8
Camberwell, Victoria 3124
Australia

Library of Congress Cataloging-in-Publication Data

Washino, Shōichi, 1945–
    Automobile electronics / Shoichi Washino.
        p. cm.—(Japanese technology reviews ; v. 1)
    ISBN 2-88124-285-5
    1.Automobiles—Electronic equipment. I.Title. II.Series
TL272.5.W37 1988
629.2'549—dc19   88-14764
                                                    CIP

Contents

Preface to the Series

Modern technology has a great impact on both industry and society. New technology is first created by pioneering work in science. Eventually, a major industry is born, and it grows to have an impact on society in general. International cooperation in science and technology is necessary and desirable as a matter of public policy. As development progresses, international cooperation changes to international competition, and competition further accelerates technological progress.

Japan is in a very competitive position relative to other developed countries in many high technology fields. In some fields, Japan is in a leading position; for example, manufacturing technology and microelectronics, especially semiconductor LSI's and optoelectronic devices. Japanese industries lead in the application of new materials such as composites and fine ceramics: although many of these new materials were first developed in the United States and Europe. The U.S., Europe, and Japan are working intensively, both competitively and cooperatively, on the research and development of high technology superconductors. Computers and communications are now a combined field that plays a key role in the present and future of human society. In the next century, biotechnology will grow, and it may become a major segment of industry. While Japan does not play a major role in all areas of biotechnology, in some areas such as fermentation (the traditional technology for making "sake"), Japanese research is of primary importance.

Today, tracking Japanese progress in high technology areas is both a necessary and rewarding process. Japanese academic institutions are very active; consequently, their results are published in scientific and technical journals and are presented at numerous meetings where more than 20,000 technical papers are presented orally every year. However, due principally to the language barrier, the results of academic research in Japan are not well-known overseas. Many in the U.S. and in Europe are thus surprised by the sudden appearance of Japanese high technology products. The products are admired and enjoyed, but some are astonished at how suddenly these products appear.

With the series *Japanese Technology Reviews,* we present state-of-the-art Japanese technology in five fields:

> Electronics,
>
> Computers and Communications,
>
> Manufacturing Technology,
>
> New Materials, and
>
> Biotechnology.

Each tract deals with one topic within each of these five fields and reviews both the present status and future prospects of the technology, mainly as seen from the Japanese perspective. Each author is an outstanding scientist or engineer actively engaged in relevant research and development.

We are confident that this series will not only give a bright and deep insight into Japanese technology but also be useful for developing new technology of our readers' own concern.

As editor in chief, I would like to acknowledge with sincere thanks the members of the editorial board and the authors for their contributions to this series.

TOSHIAKI IKOMA

Preface

Recently, Japanese automotive technology has accomplished remarkable progress, and many automobiles are produced in Japan (e.g., 7.8 million passenger cars in 1986). This is attributable to the development of mass production technology with both high reliability and quality control, and the development of control technology in the automobile industry. It seems that basic research on the engine, the body, the materials, control and so forth have brought about such developments. However, research is often not published in English. This arises from both the language difficulty and lack of communication due to the distance between Japan and Europe and the U.S.A. This book is an introduction to state-of-the-art Japanese automotive technology, focussing mainly on automotive control.

First, a brief overview of automobile control technology in Japan is summarized and the concept of the intelligently controlled vehicle (ICV) being developed in Japan is introduced. Several steps toward commercialization of the intelligently controlled vehicle and related problems are pointed out. Then, models for both the idle state and fuel transportation useful to the first step toward the commercial production of ICV are introduced. It is shown that the models can also explain the phenomena we often meet in actual control. For example, the analysis both on the ISC system and the idle stability are shown.

Finally, the author introduces the status of the data-driven processor, which is probably useful in commercialization of the ICV.

Literature published in Japan are listed at the end of the booklet.

Acknowledgments

I wish to acknowledge Dr. Harashima, Dr. R. Ohnishi, and Dr. Yano for allowing me to accomplish this work, and also to recognize the contributions of R. Nishiyama, S. Ohkubo, and T. Inoue (especially, Chapters 1 and 2). The contributions of K. Shima, M. Meichi, T. Fukuhara, T. Yamasaki, and S. Komori are also gratefully accepted.

SHOICHI WASHINO

Nomenclature

Chapter 1

A/F	air fuel ratio
J	cost function
SA	spark timing
EGR	exhaust gas recirculation
T	time required for driving regulated driving mode
$C^*(t)$	admissible cyclic torque variation
$C(U,t)$	instantaneous torque variation
t	time
W	weight in cost function
X_i	accumulative emission
H	Hamiltonian
p_i	Lagrange multiplier
f_i	time derivative of p_i
z	z-transformation
$x(k)$	state variable of system
$y(k)$	controlled output
A, B, C	system matrices
$U(k)$	control input

Chapter 2

$\dot{G}_a$	air flow rate through throttle plate

$\dot{G}_e$	air flow rate through intake valve
G_m	air weight in intake manifold volume
J	inertia moment of flywheel
L	dead time
ω_n	natural frequency
ζ	resistance coefficient
K_c	proportional gain in ISC controller
T_i	integral time in ISC controller
T^*	characteristic time constant of engine in idle
$S=jw$	imaginary unit
Subscript	
Δ	means variation from equilibrium
o	means equilibrium value

Chapter 3

τ_d	dead time
T_w	manifold wall temperature
$\dot{G}_i$	injected fuel flow rate
$\dot{G}_v$	vaporized fuel flow rate
G_l	weight of fuel film
M_g	molecular weight of fuel
A_s	effective area of fuel film
P_s	saturated vapor pressure
D_v	diffusion coefficient
S_h	Sherwood number
T_a	air temperature in manifold
C_{pl}	specific heat of fuel
P_b	manifold absolute pressure (MAP)

V_m	manifold volume enclosed by inspection surface
R	gas constant
T_m	temperature of air in intake manifold
N	engine speed (RPM)
η_v	volumetric efficiency
τ_a	time constant of manifold system
V_h	stroke volume of engine (litre)
P_w	injection pulse width
τ_q	dead time due to stroke delay
K_d	coefficient of derivative fuel modulation
T_{inj}	interval of fuel injection
$M(S)$	modulation function of fuel
K_v	rate of fuel carried by air flow to that of injected fuel
τ_m	time constant of fuel transportation
$G_n(S), G_f(S), G_b(S), G_c(S),$	transfer functions
T_b	total brake torque
K_p	product of P_{bo} and derivative of T_b about P_b
K_λ	product of λ_o and derivative of T_b about A/F
K_n	product of N_o and derivative of T_b about N
T_l	temperature of fuel film
D_l	effective droplet diameter (experimentally determined)
λ_l	thermal conductivity of fuel film
A_c	fuel film area
T_w	wall temperature
H_l	latent heat of fuel film
T_f	temperature of injected fuel
δ	density of fuel film
R_e	Reynolds number

ν	kinematic viscosity
T_i	indicated torque
P_i	indicated mean effective pressure
T_{fr}	friction torque
P_f	friction mean effective pressure
$\phi(\lambda)$	coefficient of indicated mean effective pressure depending air ratio
$K(\epsilon)$	coefficient of indicated mean effective pressure depending
ϵ	compression ratio
t_c	intake air temperature
P_r	back pressure

CHAPTER 1

A Brief Overview of Electronic Control in Automobiles

A Proposal for an Intelligently Controlled Vehicle

ABSTRACT

In this chapter, a brief overview of the status in Japan of electronic control in automobiles, especially engines, is presented. Then, by considering the status, the author infers the future prospects of electronic control in automobiles, and proposes an idea for an intelligentlly controlled vehicle (ICV) as an R&D concept. And, finally, several problems for commercialization of the concept are stated.

1. Introduction

Since the regulation of exhaust gas emissions, electronic control has been replacing mechanical control. A large number of electronic engine control systems are now installed. For example, electronic fuel injection, spark timing or knocking control, and idle speed control are noteworthy. Recently, other electronic control systems for both the transmission and the vehicle, such as 4WD, 4WS, and ABS, have been commercialized.

These replacements are brought about by both social requirements, such as clean exhaust gas and the remarkable progress of semiconductor devices and their high versatility. They will continue to progress gradually. In the future, there will be new era for both automobiles and electronic control systems in automobiles[1][2]. A main purpose of this chapter is to make clear what the new age is: a brief overview of the status of electronic control in automobiles is described together with several problems and the steps toward commercializing future control.

2. Status of the electronic control in automobiles

2.1. General view of electronic control in automobiles

Representative electronic control systems and their features are summarized in Table 1[3]. Since more details of these control systems are explained in the proceedings of the symposium sponsored by Electronics Committee of Japanese Society of Automotive Engineers(JSAE), brief explanations are presented below[4]-[6].

In Table 1, the systems to control the injected fuel quantity, the

Table 1. Control in automobiles.

Name	Controlled variable	Control input	Control logic	Detected variables and actuator
Fuel control	Air/Fuel ratio	Injected fuel	Optimal control (Open)	Air flow Fuel injector
EGR control	EGR rate	Valve opening	ditto	Valve position EGR valve
Spark timing control	Spark timing	Primary current	ditto	Crank angle Oscillation of cylinder pressure
Idle speed control	Idle speed	Air flow rate (current)	PI LQI	Engine speed ISC valve
Cruise control	Vehicle speed	ditto	PI control	Vehicle speed Throttle actuator
Transmission	Gear ratio	Hydraulic pressure (current)	Open loop	Vehicle speed MAP Control valve
4WD	Torque distribution ratio	ditto	ditto	Engine speed Steering wheel angle Control valve
4WS	Wheel angle	Stepping motor rev. angle	PI Open	Vehicle speed Wheel angle Stepping motor
ABS	Slip ratio	Hydraulic, pressure (current)	PI LQI control	Vehicle speed Wheel speed Control valve

EGR rate, and the spark timing were commercialized in the earliest period to optimize the fuel cosumption under stringent CO, HC, and NOx emissions regulations. One of the purposes of fuel control is to keep the air/fuel ratio(A/F) at the stoichiometric ratio by injecting the fuel proportional to the measured air flow rate through the fuel injector. And, as a result, the poisonous components in the emission gas are reduced by operating the 3-Way catalyzer effectively. The EGR control is used for the reduction of the NOx emission by recirculating the exhaust gas from the exhaust manifold to the intake manifold because it decreases the flame temperature in the cylinder. The maximum torque of the engine is obtained by controlling precisely the spark timing according to the operating point of the engine.

A typical construction of these electronic control systems is shown schematically in Figure 1. The important sensors used in the controls are the air flow sensor, the crank angle sensor, the knock sensor, etc. The fuel injector and the EGR valve with a position sensor are used as the important actuators. Recently, the cylinder pressure sensor, which detects the cylinder pressure oscillation directly, tends to be adopted instead of the knock sensor, which detects the oscillation indirectly through the oscillation of the engine block that is induced by cylinder pressure oscillations.

Generally speaking, the better combustion becomes in the cylinder, the more the poisonous gas (NOx) increases and the the fuel consumption becomes less. Therefore, it requires a certain optimization technique for us to reduce the fuel consumption under emission regulation. The control algorithm used there is based on an optimization control theory such as the maximum principle or dynamic programming. Although the applicability of these theories to the controls of fuel, EGR, and spark timing was proposed and discussed in the USA by Cassidy[7][8] and Dohner[9][10] now also in Japan, the optimization technique is widely used as a basic method of engine calibration.[11][12] In the actual engine calibration, however, it is found that the method still needs many processes to harmonize between good drivability, low fuel consumption and the reduction of poisonous gas, because the cost function of the drivability is not formulated now, as shown in the next section.

If the formulation of the drivability were completed, the engine

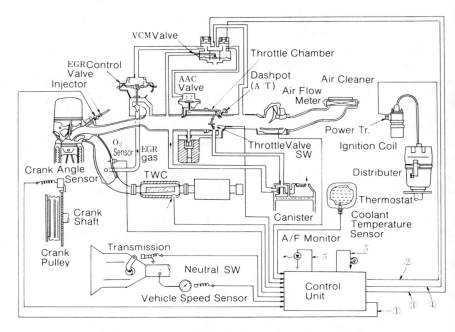

Figure 1. An example of electronic engine control

calibration processes would be simplified remarkably. Moreover, if it is implemented in the on-board computer, the processes may be carried out automatically by the vehicle itself. This is a kind of the self-turning and may be one of the future aspects of the fuel, EGR, and spark-timing controls.

In the actual control, the predetermined values are stored in the memory in the microprocessor, and they are used as the target values of the controlled variables. They, therefore, cannot be altered once they are determined in the present control as will be stated later.

The idle-speed control (ISC) in Table 1 belongs to one of the engine controls and its purpose is to maintain the idle engine speed at the desireable level independently of several torque disturbances by controlling the air-flow rate through either the throttle valve or the ISC valve in the bypass passage. The schematic diagrams of the

ISC valve are listed in Table 2.[13] The bypass type ISC is widely used now because it can control the air flow rate more easily than the throttle type because of its moderate sensitivity against the the air-flow rate. That is, in the throttle type the air-flow rate changes remarkably against the slight change of throttle opening, and, there-fore, it needs more accurate control of the throttle opening. The detection of the engine speed is realized by measuring the ignition interval by the microprocessor.

As shown in the next section in detail, the PI control theory is still widely used as the control algorithm. As shown in later chap-ters, the idle state of engines is a very unstable system from the point of view of control theory. Therefore, control results against the torque disturbance are often unsatisfying. For example, slow restoration of the idle engine speed under a torque disturbance is experienced frequently. For improving this, a kind of the feed-forward technique is often used. For example, in order to treat the torque disturbance of the air-conditioner as a known disturbance, the switching signals of the conditioner are often lead to the com-puter input port. And the control action starts before the torque disturbance is actually applied. As a result, this prevents the abnor-mal decrease of the idle speed to some degree.

Recently a trial for improving the control result by using modern control theory was proposed. Takahashi et al. discussed an applica-tion of the LQI (Linear Quadratic Integral) control to the idle-speed control and showed that the control result was improved to some extent.[14] However, it seems that such a response as the abnormal decrease of engine speed against the load disturbance is not greatly improved. The reason for this would be due to the lack of accurate modelling of the idle state. If we can get a more accu-rate model the response will be improved greatly, and such control gains as P and I gains would also be self-turned. From this point of view a more accurate model is proposed in the later chapter.

The transmission control shown in Table 1 is used to make the gear ratio of the automatic transmission optimal by controlling the hydraulic pressure according to the operating point of the engine. The optimal gear ratios predetermined by various experiments are listed in the data table of the the microprocessor memory against the operating points of the engine. And, in the actual control, the

Table 2. Comparison of commercial ISC.

Air passage, Notes	ISC valve	Notes
1) Throttle valve 1. Suitable for SPI 2. High sensitivity for air control 3. Free from contamination 4. Large package	1.DC Motor Gear DC Motor	1. Small package and Light weight 2. Current required during operation 3. Position sensor required
	2.Vacuum Motor Atmosphere	1. Various components 2. Large package 3. Current required during operation 4. Position sensor required
2) Bypass passage 1. Low sensitivity for air control 2. Needs some consideration against contamination 3. Small package 4. Easy installation	3.DC Motor Gear DC Motor	1. Small package and Light weight 2. Current required during operation 3. Position sensor required
	4.Stepping Motor Stepping Motor	1. Brushless (long life) 2. Current required during operation 3. No position feedback is necessary
	5.Linear Solenoid	1. Simple construction 2. Current always required 3. No position feedback is necessary 4. Hystelesis and small error involved
	6.Vacuum Motor Atmosphere	ditto. except 4

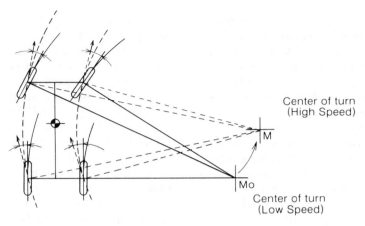

Figure 2. Relation between the center of turn and turning velocity

gear ratio is controlled so as to agree with the data written in the memory. This means that the aimed values of the control written in the memory are not changed on board against the deteriorative variations of the mechanical components once the aimed values are predetermined. Therefore, after a cetain mileage the gear ratio is not necessarilly controlled at its optimal value. This is one of the greatest drawbacks of the present transmission control as well as of the engine control.

The 4-Wheels Drive (4WD) control is used to get the optimal torque transmission ratio between the front wheel and the rear wheel. It is well-known that it results in the optimal torque transmission ratio not only in braking but also in accelerating. Therefore, the study about this was started very early, and the mechanical control system of 4WD has been developed. One of the most difficult points of 4WD control is that it is necessary to slide the transmission torque between the front and the rear wheels according to the turning speed of the vehicle from the point of the stable driving. This situation is explained in Figure 2. When the vehicle turns slowly, the center of turn coincides with the point MO, but the center approaches the point M with higher turning velocity of the vehicle. Therefore, if the drive shaft for the torque transmission between the front and rear wheels is connected rigidly a braking phenomenon in slow turn is caused. Because of this, it is necessary to slide the transmission torque between the front and the rear wheels according to the vehicle speed. For realizing this, the following means are considered: (A) use of the 3rd differential gear, (B) use of 4 Wheels Steering with the inverse phase between the front and the rear wheels in slow turn, and (C) change of 4WD into 2WD when we turn the vehicle slowly and (D) slide the transmission torque between them. Except for (B), the other three have been realized rather easily by using the mechanical components. Thus, in 4WD as well as 4WS, little contributions of the electronic control are observed. It seems that the reliability of the electronic components is not believed fully yet in this control field, although the progress of the electronic components makes them reliable and cheap. In spite of this, some applications are found in (C) and (D), as shown in Figure 3. This example adopts the selection type (C), the selection between 4WD and 2WD electronically controlled by detecting the slip between the front and the rear drive shafts through both the angular displacement of the steering wheel and every wheel revolution. The actuator for the selection is MT-T and actuated by the hydraulic pressure activated by the computer signal.

The constant vehicular speed control (the cruise control) maintains the vehicle speed at the desirable one independently of the load resistance by the throttle actuator which alters the throttle opening. A block diagram of the cruise control is shown schemati-

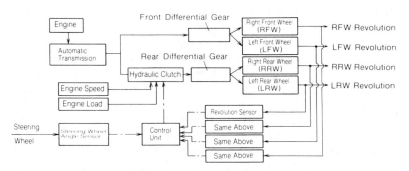

Figure 3. An example of 4WD with the selection type

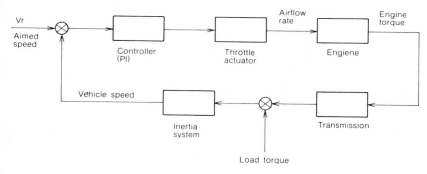

Figure 4. A block diagram of cruise control

cally in Figure 4. PI control theory is widely used as the control algorithm, but the satisfactory control results are not always obtained especially in the transient response. The same situation is ISC is considered, because the control object in this case would also have second-order lag. Therefore, the large phase lag prevents the controller from having large PI gains. Thus, the transient properties of the control and the stability of the control system are not very good. Recently, Tabe et al. discuss the application of modern control theory and demonstrate successful results.[15]

The 4 Wheel Steering control (4WS) maintains the rear steering angle at its desired level for obtaining more stable driving. For example, as shown in Figure 2 at low turning velocity of the vehicle, the steering angle of the rear wheel is controlled to the inverse phase against the front one. As a result, the vehicle can turn with a smaller turning radius. On the other hand, at higher turning velocity the steering angle is controlled to the same phase as the front wheels and this gives the same effect as at lower turning velocity. A simple control system for 4WS is given in Figure 5.

Generally speaking, with 4WS the electronic control system is not used so much as in 4WD, because misoperation of the system will cause extreme damage for the driver. This suggests that the establishment of appropriate measures against failures of the electronic control system is greatly important. This is also the case with other electronic control systems in automobiles.

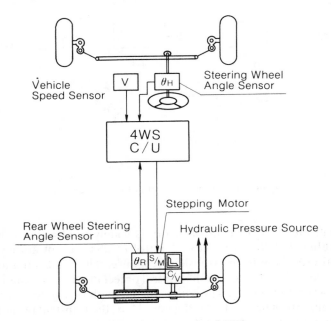

Figure 5. An example of 4WS

The Anti-Blocking Brake System (ABS) keeps the slip ratio between the tire and the road optimal, e.g., 20%, and prevents the lock phenomenon by controlling the hydraulic pressure of the braking system according to the slip ratio. A typical system construction is shown in Figure 6. Also, in this case, PI control theory is widely adopted.

2.2. Overview of the control technology

As is well-known, sensors, actuators, microprocessors and control theory are indispensable for the control of automobiles. All of these factors considered together may be called control technology. In this section, a brief overview of control technology in automobiles, especially in engine control will be presented.

First, the characteristics of the three controlled objects given in

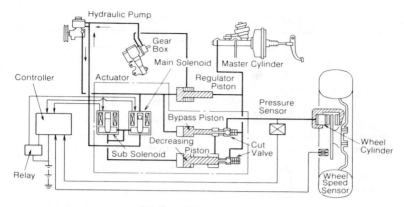

Figure 6. An example of ABS

Table 1 are shown in Table 3.[3] From this table, one can see easily that the engine is one of the most difficult objects to control because it has the characteristics of nonlinearlity, high response, large variation of parameters, multiple inputs and output that are mutually dependent, and there is dead time. These characteristics make it difficult for us to make an accurate modelling of the engine. On the other hand, the characteristics of the power train and the dynamics of the vehicle are a bit simpler, although the parameter values of these objects also vary greatly depending on the operating points. However, there is a difficulty in formulating the cost function of human feelings such as drivability and shock in changing the gear ratio. These are barriers for the application of advanced control theory.

Now, we overview briefly the sensor, the actuator, and the control logic used in the actual control. It is obvious that the key sensor in the engine control which includes the fuel, EGR, and the spark timing is the air-flow meter. The characteristics of the air-flow meters used practically are listed in Table 4.[16] As is well-known, the manifold absolute pressure (MAP) sensor measures the air-flow rate indirectly and all the others directly. The MAP sensor has advantages such as compactness, cheapness, large freedom of installation, and no pressure drop. However, it also has the

Table 3. Features of controlled objects.

Objects	General features
Engine system	1. Nonlinear system 2. Fast response 3. Large parameter variation 4. Includes multi-input and multi-output system 5. Dead time system
Drive-Train system	1. Rather simple model compared with engine system 2. Difficult formulation of cost function (e.g. drivability)
Vehicle system	1. Rather simple model compared with engine system 2. Large parameter variation 3. Difficult formulation of cost function

disadvantage that the controlled air-fuel ratio varies easily with both the engine variation and the introduction of the exhaust gas recirculation (EGR), because both the valve timing and EGR alter the volumetric efficiency under the same manifold pressure. The MAP sensor, therefore, is often used in cheaper control systems without EGR. On the other hand, air-flow meters such as the moving vane, Karmann vortex, and the hot wire cover the disadvantage of the MAP sensor though they do not necessarily have the advantages of the MAP sensor. Thus, air-flow meters that can measure the air-flow rate directly are widely used because of their capability for EGR. The key actuator in the engine control is the fuel injector, because air-fuel ratio control by the fuel control is the most important of all. Although both the spark-timing control and the control of the exhaust gas recirculation is also very important for achieving both the clean exhaust gas and the optimal torque, these controls become most effective when the air-fuel ratio is maintained at the stoichiometric ratio accurately.

A cross section of a fuel injector is shown in Figure 7. This injector has two sprays with good atomization and is used frequently for engines with both the double over head cam and two intake valves. The increase of the dynamic range and the good atomization of the injected fuel are the technical points to be improved for future injectors.

The optimization control logic with brief explanations are shown

Table 4. Air flow meters practically used.

	MAP sensor	Moving vane	Karmann vortex	Hot wire
Output signal	Electric potential proportional to MAP (Analog)	Electric potential proportional to dynamic pressure (Analog)	Frequency proportional to volume flow (Digital)	Electric potential proportional to mass flow (Analog)
Accuracy	3%	3%	3%	3%
Pressure drop[1]	0[2]	110	70	55
Accuracy in WOT[3]		18	9	8
Advantages	Small package Cheap Installation adaptability	Long record of performance	Easy signal processing due to digital output Good durability	Detectability of mass flow
Disadvantages	Low adaptability to EGR and engine variation	Large package Degradation of moving parts	Atmosphere compensation is needed	Degradation due to comtamination Large package

Notes 1. Pressure drop (mmAq) at 50l/s 1mmAq = 10Pa
    2. MAP sensor detects manifold absolute pressure
    3. Error rate of output when MAP = 750mmHg

in Table 5. The formulation of the optimization process is mentioned below.[9][10]

Assuming that the control input $U = U(A/F, SA, EGR)$, the cost function J can be written by Eq. (1-1) and the constraint by Eq. (1-2).

$$ J = \int_0^T \left[L(U, t) + \frac{C^*(t)}{W\{C(U, t) - 2 C^*(t)\}} \right] dt \qquad (1-1) $$

$$ X_1(T) < HC^* \quad X_2(T) < CO^* \quad X_3(T) < NO_x^* \qquad (1-2) $$

where A/F, SA, EGR describe the air fuel ratio, spark timing, EGR ratio, respectively. Also, L is the fuel flow rate, T the time

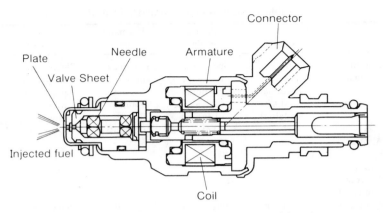

Figure 7. A cross section of the fuel injector with two sprays

interval of the driving mode, C^* (t) the admissible cyclic torque variation, W the weight, $C(U,t)$ the instantaneous torque variation, and X_i the accumulative emission.

The problem is reduced to finding the control input value of U that minimizes J and satisfies Eq. (1-2) under a given driving mode such as shown in Figure 8. Thus, our task is to derive the value of U which minimizes the following Hamiltonian which includes the Lagrange multiplier p.

$$H = L\ (U,\ t) + \frac{C^*\ (t)}{W\{C\ (U,\ t) - 2C^*\ (t)\}} + \sum_{i=1}^{3} p_i f_i \quad (1-3)$$

If a simple and analytical engine model is obtained, we can differentiate Eq. (1-3) and obtain easily the optimized control input value U_{opt}. The operation of engines is very complex, and, hence, simple models for engine characteristics cannot be obtained. Therefore, as shown in Table 5, various efforts have been made to solve this problem. For example, A.R.Dohner proposed the gradient iteration method, and Matsumoto et al.[11] used orthogonal polynomials representing the engine characteristics and determined U_{opt} by the Lagrange Multiplier method. Moreover, Ikeura et al.[12] determined

Table 5. Proposed optimization techniques.

No.	Classification	Literature	Abstract
1	Maximum-Principle or Lagrange-Multiplier	A.R. Dohner SAE 780286	Gradient iteration method: Seeks Uopt by iteration till perturbed H < admissible value. Drivability function is defined by variance of IMEP.
2		J. F. Cassidy SAE 770078	Perturbed Lagrange Multiplier: Seeks Uopt by iteration till fuel consumption is minimized under emission constraint.
3		Matsumoto et al. SAE 780590	Lagrange Multiplier: Regression model using orthogonal polynomial about A/F, EGR, N, T etc. Parameters are determined by least-squares method.
4	Linear-Programming	E. A. Rishavy et al. SAE 770075	Linear programming
5	Dynamic-Programming	T.Trella SAE 790179	Dynamic programming: Regression model using second order polynomial about A/F, EGR, N, and BMEP
6		Y. Hata et al.	Dynamic programming
7	Non-linear Programming	H. Rao et al. SAE 790177	Non-linear programming
8	Maximum-Principle	J.F. Cassidy IEEE Trans. AC-25-5 (1980)	Theoretical discussion by using LQ control: Linearized mathematical engine model. Pade approximation of dead times. Stability analyses.
		J.B. Lewis GM Report CR-80-12/ET	Much the same as above. Parameter sensitivity analysis and consideration of Gaussian random noise.

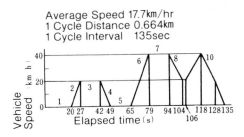

Figure 8. Japanese 10-modes driving pattern

by dynamic programming. Generally, such calculations are carried out by supercomputers because of the immense calculation processes involved.

The predetermined U_{opt} by the process mentioned above is written into the read only memory (ROM) of the microprocessor and used as the target value of the actual control, as mentioned previously. Thus, if the optimal control input values are not renewed adaptively according to the changes in the engine characteristics with the mileage, the control deviates from the optimal control. Although this is one of the disadvantages in the present control technology, optimal control theory is practically used in the form mentioned above.

We review next the actuator and the control logic practically used in ISC. Generally, ISC is the most important control in the engine control, because the time variations of the idle speed often cause various unpleasant oscillations for the driver and the engine stalls in the worst case. On the other hand, the stability of combustion at idling improves the fuel consumption. Thus, it is believed that controlling idling is controlling the engine.

Concerning the actuator that adjusts the air-flow rate through either the throttle valve or the bypass passage (ISC valve), explanations are given in Table 2. Explanations on the control logic used in ISC will be described below. As mentioned earlier, the PI feedback control theory widely used at present does not always give satisfactory control results. Thus, it seems worth reviewing the discussion on the applicability of the LQI control to ISC proposed by Takahashi et al.[14]

An example of the physical construction together with a block

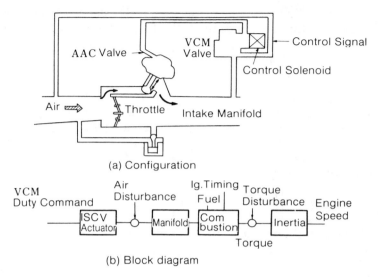

(a) Configuration

(b) Block diagram

Figure 9. Both physical configuration and a brief block diagram of ISC

diagram is given in Figure 9. The ISC valve of this example resembles the vacuum motor in Table 2; the air-flow rate through the bypass passage is adjusted by controlling the opening of the AAC valve and the drive of the valve is carried out by the control pressure in the VCM valve. The control pressure consisting of the atmospheric and the manifold pressure is controlled by the duty ratio of the control signal applied to the control solenid. The system block diagram from the duty command to the engine speed is given in Figure 9 (b).

As is easily understandable, the procedure for LQI controller design lies in: (1) identification of the transfer function from the duty signal to the engine speed, (2) derivation of both the system equation in the form of the matrix and the cost function, and (3) obtaining the optimal gains by solving the Riccati equation in matrix form. Takahashi et al. identified the transfer function by the least-squares method and obtained Eq. (1-4).

$$T(z) = \frac{0.222z + 4.167}{z^2 - 1.655z + 0.719} \tag{1-4}$$

Then they gave the system equation as the minimal realization expressed by Eqs. (1-5) and (1-6) in the discrete form from Eq. (1-4).

$$x (k + 1) = A^* x (k) + B^* u (k) \qquad (1\text{-}5)$$

$$y (k + 1) = C^* x (k) \qquad (1\text{-}6)$$

where, x is the two-dimensional state variable, and the second element in x is the engine speed. The first element is a quantity without a physical meaning and is evaluated by the state observer shown in Figure 10. The quantity u represents the duty command as the control input, and the coefficients $A,^* B,^*$ and C^* are given by the following equations.

$$A^* = \begin{bmatrix} 0 & -0.719 \\ 1 & 1.655 \end{bmatrix} B^* = \begin{bmatrix} 4.167 \\ 0.222 \end{bmatrix} C^* = [0 \ 1]$$

The cost function is given by the following equation.

$$J = \Sigma[e^2 (k) + \alpha W^2 (k)] \qquad (1\text{-}7)$$

where

$$e(k) = y_r - y(k) \qquad W(k) = U(k) - U(k + 1) = \Delta U(k)$$

Rewriting Eq. (1-5) and Eq. (1-6) in terms of $\Delta U(k)$ and solving the Riccati equation in the matrix form, the optimal feedback gains

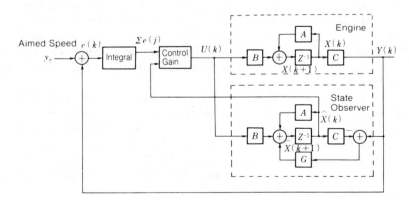

Figure 10. A block diagram of ISC with LQI control

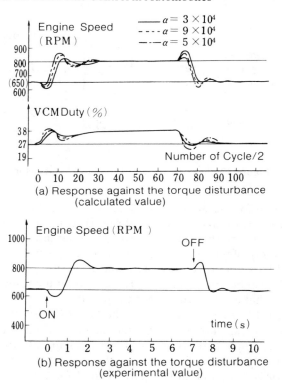

Figure 11. Examples of the idle-speed control with LQI control

shown in Figure 10 are obtained. A comparison between the calculated and the experimental control result is given in Figure 11. The
parameter α is the weight shown in Eq. (1-7) and the smaller it
becomes the less the variation $W(k)$ (VCM duty), which is the
reasonable tendency. It is reported that the experimental results to
prevent the torque diturbance of air-conditioner showed improvement compared with the conventional PI control.

Recently, other studies on the applicability of modern control
theory are reported. For example, Tabe and Ohba et al.[17] reported
the applicability of the LQI control to the ABS control shown in
Table 1, the temperature control in the room, and the control of the in-

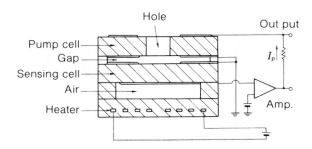

Figure 12. A cross section of the linear O_2 sensor

jection pump for diesel engines. Moreover, H. Takahashi showed the applicability of fuzzy control to the cruise control.[18] Though we have not yet heard that these controls were put into practical use, the importance of their studies remains to be appreciated.

Several barriers preventing the application of the modern control theory will be discussed later. Here we describe some interesting sensors that have recently been proposed.

In Figure 12, the cross section of the linear O_2 sensor is shown, which gives linear outputs against the air-fuel ratio in the exhaust gas.[19] An example of the output is shown in Figure 13, which shows outputs greatly different from those of the conventional λO_2 sensor. This sensor is composed of two parts made from ZrO_2: one part is called the pumping cell, which pumps out the O_2 molecules from the gap to the outside, and the other is called the sensing cell, which detects the electromotive force between the reference and the gap.

The pumping current is controlled so that the voltage is equal to the reference voltage (e.g., 0.4V). Accordingly, the current becomes nearly proportional to the air-fuel ratio in the exhaust gas. Thus, it is expected that the ratio can be controlled at an arbitrary value and that the response of the A/F control becomes fast compared with that obtainable with the λO_2 sensor, because the error signal between the aimed value and the actual A/F value can be detected.

The cylinder pressure sensors shown in Figure 14 and 15 are also proposed for applications to the knock control and the A/F control. Hata et al.[20] showed the applicability of a piezoelectric type sensor

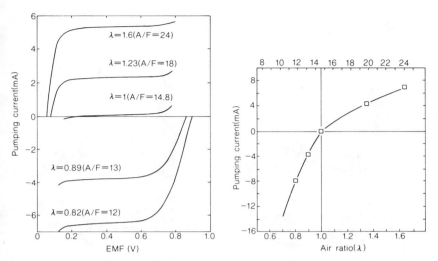

Figure 13. Examples of the output characteristics of the linear O_2 sensor

shown in Photo. 1 both to the knock control[21] and to the A/F control. Sasayama et al.[22] proposed an optical pressure sensor with a silicon diaphram shown in Figure 15 for engine control. These sensors seem promising for future applications in automotive engine control.

3. Some considerations on future control

We discuss here the future aspects of the automotive control based on the present situations described above and the potential needs for automotive control. In Table 6, the potential needs proposed by control engineers and users are summarized. One of the most earnest needs pointed out by engineers would be the autotuning controller. If the control system on board can determine either the control gains or the aimed values of the control by itself, engineers only have to distribute the necessary components for the control and connect them. Then the optimization of the control is automatically carried out by driving the car according to the regulated

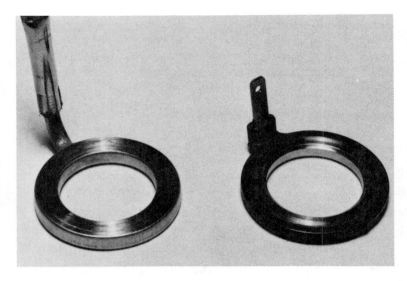

Photograph 1. Examples of cylinder-pressure sensor

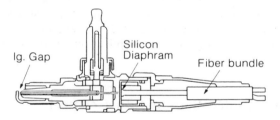

Figure 15. An example of the optical cylinder pressure sensor

driving mode. What a splended thing this is! As stated above, under the present situation, the control engineer must spare his time for the determination of the control gains or the target values. What they have to do is to collect an immense amount of experimental data from the engine bench and actual driving and to determine the gains by trial and errors.

The gains predetermined by such laborious efforts become not optimal with the mileage, due to the gradual change of the me-

Table 6. Needs for automobiles.

User	Engineer	Social
Control	Self-tuning control	Environmental
1) Meeting drivers	1) Optimal control	Pollution
preference	2) Intelligent learning control	1) Low emission
2) Fast response	3) Adaptive control	2) Low noise
Engine	4) Automatic calibration	3) Traffic
1) High power	Precise control	control
2) High smoothness	1) Precise control results	Energy
3) Low noise	2) Direct sensing of variables	1) Low fuel
4) Low fuel	Total control	consumption
consumption	1) Drive by wire	
	2) Local area network	

chanical characteristics. One of the purposes of the attempt to apply the modern control theory to the control system was to reduce the disagreeable labor, because theory taught the theoretical determination of the gains. However, it seems that the attempt does not always succeed. For these reasons, we remember easily the items such as the poor robustness as shown in Table 7. More detailed discussions will be given later. Precise control, intelligent learning control, adaptive control and the low emission in Table 6 are essentially the same as self-tuning, because these function are realized by the self-tuning function.

Is self-tuning favorable only for control engineers? No, it is favorable also for users, because it satisfies such needs of users as shown in Table 6. Lately, the preferences of users have diversified pretty much according to their values. For instance, some users may be satisfied with a vehicle which runs automatically, without their command or driving, but some may demand further that the vehicle should run by itself following his intention or feeling. The feeling of a driver would be different every day. Sometimes he thinks that the car should run following his thought and sometimes he probably expects that his car should run without his command driving. Sometimes he would want to drive fast or slow. Moreover, a beginner driver would think that the car should assist his driving. All in all, it is desirable to change the control according to the intention of the

Table 7. Comparison between modern and classical control theory.

Item	Modern control theory	Classical control theory
Knowledge about object	Concrete knowledge (including sensors and actuators)	Non-concrete knowledge
Knowledge for designing controller	Linear algebra; Simulation (Time domain)	Complex variable analysis (Frequency domain)
Designing for multi-variable system	Same as single variable system	CAD for multivariable system
Designing for system with dead time	Pade approximation	Smith method
Robustness of controller	Poor; Guaranteed in special case	Guaranteed with large controller gain
Optimization	Easy	Difficult
Empirical factor in designing controller	Weight in cost function	Determination of controller gain

driver. Such control could be realized, if the control system were able to understand the drivers intention through a man-machine interface. The vehicle with such control may be called an intelligently controlled vehicle (ICV). It is favorable not only for engineers and users but also for car makers, because car makers are interested in new models with a few different specifications. Thus, the standardization of mass production is realized in spite of the various preferences of users. Several car makers in Japan have also demonstrated concepts very close to ICV in the International Tokyo Motor Show 87.[23] An example of the concept is shown in Figure 16. Therefore, this concept of ICV is not a personal idea but it will become common in Japan.

It is doubtless that the fail-safe mechanism shown in Table 6 is also important for engineers and drivers. In the present situation, the concept of fail-safe mechanism is considered in the individual devices such as actuators and microprocessors. However, it seems that a systematic concept of fail-safe for automobiles is not yet established. One reason may be the cost. In flight control systems,

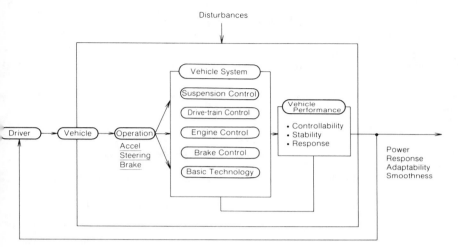

Figure 16. A concept resembling intelligently controlled vehicle (at Tokyo Motor Show)

function is more important·than cost and, therefore, one can pursue the best function without consideration for the cost. But in an automotive control system, the cost is very important as well as the function. Therefore, in the automotive control system, a new fail-safe concept that satisfies both the function and the cost at the same time is necessary. As shown later, in ICV, communication among all control systems is carried out mutually and reflected in the control action based upon the information from other control systems. In such a situation, fault tolerant theory may be useful to the realization. This could give an index to the solution of the problem.

Table 8 shows the technologies and steps necessary for the commercialization of ICV. The first step, as mentioned above, is to provide the control systems with the self-tuning function. The function of understanding the intention of the driver is the second step. The third step is to provide the systems with the communication function in the systems. And, finally, the control systems have to be made capable of using external information to control the vehicle.

Based upon Table 8, we review the technology necessary to

Table 8. Approaches to ICV.

Step	Point of development		Remarks
First step	Self-tuning controller		
	1) Sensors	Exhaust gas sensor	NOx, CO, HC
		Fuel flow sensor	
		Drivability sensor	e.g. G-sensor
		Cylinder pressure sensor	Absolute value
		Torque sensor	Brake torque
	2) Control logic	Advanced control theory	Table 10
	3) Modelling	Exact model on subsystem	e.g. Idle state Combustion
	4) Processor	Non-von Neumann architecture	e.g. Data-driven processor
	5) Cost function		Time or frequency domain
	Fail safe	Should satisfy both cost and performance	Component System
Second step	Detection of drivers intention		
	1) Sensors	Steering wheel angle sensor	
		Position sensor of accel pedal	
		Position sensor of brake pedal	
	2) Detection logic		Knowledge about human science
Third step	Total control		
	1) Communication in vehicle	Communication between control systems	Drive by wire LAN
Fourth step	Global control		
	1) Communication with outside	Communication with station, world, etc.	e.g. GPSS

realize ICV. As for the sensors, exhaust gas sensors such as NOx, HC, and CO sensors are required: the NOx sensor is especially important. A fuel flow sensor is also required. It would require no explanation that if these sensors were developd the optimization for the emission regulation would become very easy on board. So

Table 9. Physical quantities obtained from cylinder pressure.

Detected quantity	Description of detected quantity	General feature Example of application
$P(\theta)$	Cylinder pressure at every crank angle	No calibration needed Knock control
$dP(\theta)/d\theta$	Derivative of cylinder pressure about crank angle	Easy detection No calibration needed Possibility of detecting A/F
P_{max}	Maximum value of cylinder pressure	Easy detection Drift of signal
θP_{max}	Crank angle at which P_{max} appear	Easy detection No calibration needed A/F control
σP_{max}	Variance of P_{max}	ditto
P_i	Indicated mean effective pressure	Precision measurement required Possibility of detecting degradation of engine under constant fuel composition and the altitude
σP_i	Variance of P_i	Precision measurement required Possibility of detecting A/F
P_i/P_b	Ratio of P_i to MAP	Precision measurement required Possibility of detecting indicated fuel consumption
T_{cyl}	Temperature of working gas	Estimation of No_x emission Measurement of mass needed
$dQ/d\theta$	Derivative of heat release about crank angle	Precision measurement required Measurement range between compression and combustion stroke
Q	Net heat release in combustion stroke	ditto Possibility of detecting degradation of engine under constant fuel composition

far, we have no available sensors to measure the drivability quantitatively. However, the three-dimensional acceleration sensor (G sensor) may be a promising candidate. It seems that the development of the G sensor is earnestly carried out in Japan. If we can quantify the drivability using this sensor, we can easily optimize the control of the vehicle including the drivability. As the sensor that detects the gradual degradation of the engine, both the cylinder pressure sensor and the torque sensor are considered. The former can detect the degradation of the engine as shown in Table 9 under constant fuel composition, and the latter can measure the variation of friction loss.

Now, we discuss the control logic for ICV. For this kind of logic, modern control theory is easily remembered. However, it has some weak points, shown in Table 6. For example, it has poor robustness. That is, the optimal gain determined by modern control theory is very sensitive to the variation of the state of the system, and, therefore, it is apt to become not optimum and cause miserable control results, especially against the parameter disturbances. Modern control theory is very effective only when we can obtain an exact model of the controlled objects. Only when all feedbacks of the state variables are carried out by sensors that detect the state variables robustness is theoretically guaranteed by the circular condition.[24] However, in the practical use of the theory, we often meet with the situation in which we cannot get an exact model of the controlled objects due to its complexity. All feedbacks of the state variables are not always carried out by the appropriate sensors, because some of the state variables have no physical meanings. Moreover, even if we have an exact model of the system, the parameters used in the model are generally changed by a change of the operating point. In this context, the example of ISC mentioned in the preceding section is really suggestive: the first component of the state variable $x(k)$ has no physical meaning. And, the poles of the transfer function given by Eq. (1-4) displace their position according to the air fuel ratio as shown in Figure 17. This fact points out that it is dangerous to rely only on the modern control theory.

What is the suitable control theory to overcome this situation? Essentially, the weakness of modern control theory lies in the low control gain. For this reason, a trial to raise it higher is proposed by

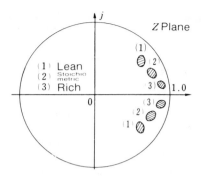

Figure 17. Examples of the pole displacements due to the variation of the aimed A/F

Fujii.[25] In my opinion, modern control theory is not always useful to ICV. Most Japanese automotive control engineers seem to have the same idea. Kuroiwa et. al.[26] summarized the useful control theories as in Table 10. It contains not only modern control theory but also the clasical control theory. Further, optimal control theory based on the frequency domain called Minimax optimizing theory is proposed. At present, however, the theory is quite difficult to understand; it may become useful to automotive engineers in the 9future. Finally, whatever the theory is, it is necessary to make full use of the appropriate control theory for the controlled objects.

Though several useful sensors are developed and some appropriate control theories are proposed, efforts to get a more exact model are always necessary. Here, it is important to note that a total model with great complexity is not always necessary. Since the development of both the sensor and the communication network will avoid the necessity of the complex model, it will be required to establish the exact submodel such as the model of the idle state, physically or mathematically. The idle model by Tkahashi et al. shown in the preceding section is certainly effective, but it cannot explain why the poles displace their position according to the air-fuel ratio. The reason for this is that the model is formulated neither physically nor mathematically. On this account, a mathematical model of the idle shown later is very interesting.

Table 10. Examples of advanced control.*

	Control method	Abstract Function	Points in adoption
Classical control theory	Non-linear PID	Adaptation of PID gains according to state of system PID with hysteresis	Identification of nonlinearity of system
	Auto-tuning	Automatic determination of PID	Determination of cost function
	Feed-Forward	Compensation before disturbance (Known disturbance)	Estimation of disturbance
	Smith method	Compensation of dead time	Estimation of dead time
	INAM	Reduces interactions between multiple inputs by diagonalization	Estimation of interactions between multiple inputs
	I-PD control	Generalization of PID Local feedback loop with PD and main feedback loop with I control	Estimation of high-frequency characteristics of system
	Minimax optimization	Optimization theory for multiple input and output system in frequency domain	
Modern control theory	Optimal regulator	Minimizes cost function with quadratic form	Determination of weight
	Pole asignment	Distributes poles of system with feedback loop according to predetermined value	Determination of desirable pole asignment
	Observer Kalmann filter	Estimates state variables from noisy signal	Determination of observer gain
	MRACS	Matches reference model to actual system by adapting parameters in reference model	Determination of reference model
	(Non) Linear Programing	Optimization of total system by (non)linear programming	Large scale system Determination of cost function
Fuzzy model	Fuzzy control	Control of very complicated system by fuzzy model	Determination of control law using fuzzy variables

*Partially edited by the author

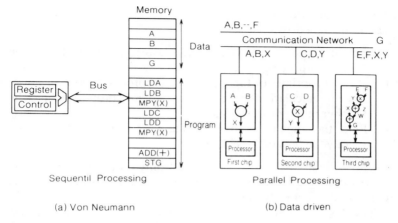

Figure 18. Processing flow in Von Neumann and data-flow architecture

On the other hand, the performance of microprocessors has to be improved greatly, since their present performance is very poor. As is well-known, the microprocessor has been progressing from 4 bits to 32 bits. Also, the research on such new materials as GaAs has been carried out actively. However, the architecture is still the Von Neumann type, and has the Von Neumann bottle neck in principle. As shown in Figure 18, both the data and the program are stored in the memory, and the central processing unit (CPU) takes the program code of the memory sequentially through the bus line. The unit fetches and decodes the code step-by-step and carries out the calculation sequentially. Therefore, when the bus line is remarkably busy, the actual operation rate of the arithmetic logical unit in the CPU is low. This situation is called the Von Neumann bottle neck, in which the calculation rate is controlled by the transfer rate both of the data and the program through the bus line. On the other hand, in an architecture of non-Von Neumann type, e.g., a data-driven type, the flow of the data drives the arithmetic logical unit and carries out the calculation. Therefore, the calculation process in the data-driven processor is represented by the data-flow graph that corresponds to the flow chart in Von

Neumann architecture often called control driven. The characteristics of the data driven processor are:

(1) Its calculation speed is generally faster than that of the Von Neumann processor under the same design rule and material.
(2) It is appropriate for the multiple microprocessor configuration, since its calculation speed is almost linearly proportional to the number of the processors used.
(3) Since the language used is diagrammatical, one can understand readily the flow of the calculation.

In Japan, many researchers study not only on the Von Neumann processors e.g., TRON[27], but also on the data-driven processors, e.g., Q-p,[28] Sigma-1[29] and DFM.[30]

The development of sensors would provide us with various new information and make it possible to realize the first step at Table 8 by making full use of both the fine control logic and the data-driven processor.

We discuss next the technology that will accomplish the second step. Apparently, the components we touch in the automobiles are the steering wheel and the acceleration and brake pedals. Therefore, information on the thought and the preference of the driver must be collected through these components; it is necessary to make clear the physical quantity that represents the driver's intention. Thus, the sensors to detect these quantities must be developed. So far, such sensors are not available. Therefore, we should also focus our efforts on the development of these sensors from now on.

Further progress of the local area network in automobile, automotive LAN so to speak, may be useful to the third step. However, for the attainment of the third step, it is also necessary to study how we utilize effectively the information from other control systems. In this sense, information theory, communication theory, fault tolerant theory, and the knowledge of systems are greatly useful. That is, the efforts at understanding the characteristics of the total system is also essential. According to the above classification, the total control including the engine, the power train, and the vehicle control will be adopted in the third step.

The final step, the so to speak global control, will have to be related to social requirements such as social traffic control. However, if the third step is completed technically there would be no anxiety. Moreover, a kind of communication system such as a navigation system and GPS is being developed, some technology of which would be useful to the attainment of the final step. Thus, this step would be accomplished rather easily compared to the preceding three steps.

4. Conclusion

A part of the present status of automotive electronics in Japan was reviewed, its future aspect was suggested, and a new concept of future automotive electronics was proposed. That is, the concept of an intelligently controlled vehicle (ICV) and the procedure for the realization of the intelligently controlled vehicle (ICV) were shown. The four steps may not necessarily be realized sequentially with the passage of time. Probably, the technologies developed individually will be coupled some day, and ICV would be realized.

CHAPTER 2

An Useful Model on Idle State and Its Application

ABSTRACT

A useful model of the idle is proposed and its applications to the analyses of idle stability and idle speed control (ISC) are carried out successfully.

A mathematical model on the idle is formulated first, and, by using the model, the stability analysis of ISC system is explained. Some measures are shown to remarkably improve the stability. Finally, it is shown that by using both an index of the negative gain margin proposed by the author and the derived model, the dependencies of the idle stability on various parameters can be explained successfully.

1. Introduction

Recently, a large number of torque diturbances has been applied to the idle, which results in abnormal transient reduction of the idle speed: in the worst case, the engine often stalls. Therefore, many automobiles equipped with the electronic fuel injection system also have an idle speed control (ISC) system to avoid the phenomenon. However, it is well-known that the control results of the ISC system are not satisfactory. Also, it is well-known that the direct cause of poor control result is low control gain. Therefore, several investigators are trying to raise the control gain. Their trials, however, seem to be unsuccessful.

In order to find out the essential reason why we cannot raise the control gain, it is necessary to get an exact or mathematical model. Many previous models of the idle were regressive. For example, F.E.Coats et. al.[31] measured the frequency characteristics of the idle and showed that the model was approximately represented by the second order lag. Takahashi et al. obtained much the same model experimentally as shown in the preceding chapter. Nishimura, Andoh, and Yamaguchi et.al. derived a mathematical model for the explanation of idle stability: the model expressed in the

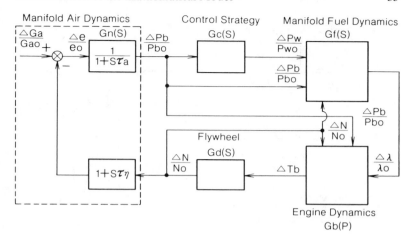

Figure 19. A linearized model of idle state

time domain was successful in the explanation of idle stability.[32][33] However, no explanation of the stability in the ISC system is given.

The author derived a mathematical model that is written in the frequency domain. Though the author's model is essentially the same as that both of Takahasi et al., and Nishimura et al., it explains not only idle stability but also the cause of the low control gain in the ISC system.

In Japan, about fifty percent of automobiles are still equipped with manual transmission. As is well-known, automobiles with manual transmission have poorer idle stability, i.e., poorer idle speed stability. Therefore, in Japan, idle stability is one of the greatest concerns. The author, however, foresees no problem with idle stability in the U.S.A.. Is this so?

2. Deriviation of the mathematical model[36]

2.1. Block diagram and basic assumptions

Figure 19 shows the block diagram of the idle. This diagram can be easily obtained by linearizing the function of every portion of the

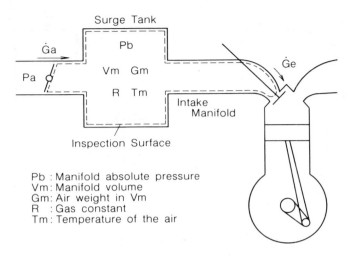

Figure 20. Manifold system

idle state in the vicinity of the equilibrium point. In Figure 19, $G_n(S)$, $G_c(S)$, $G_f(S)$, $G_b(S)$, $G_d(S)$ are the transfer functions, and the meanings of other notations are listed in the nomenclature.

2.1.1. The manifold system

The transfer function $G_n(S)$ and its related equations in the manifold system should be referred to Figure 20. The following equations are derived from the mass conservation law, the state equation of gas, and the definition of the volumetric efficiency.

$$\dot{G}_a = \dot{G}_m + \dot{G}_e \tag{2-1}$$

$$P_b V_m = G_m R T_m \tag{2-2}$$

$$\dot{G}_e = \frac{\eta_v P_b V_h}{120 R T_m} N \tag{2-3}$$

By taking the variation Δ of the various quantities in the above equations, we obtain

$$\frac{\Delta \dot{G}_a}{\dot{G}_{ao}} = (1 + S\tau_a)\frac{\Delta P_b}{P_{bo}} + \frac{\Delta N}{N_o} + \frac{\Delta \eta_v}{\eta_{vo}} \qquad (2\text{-}4)$$

and

$$\tau_a = \frac{120}{\eta_{vo}N_o}\frac{V_m}{V_h} \qquad (2\text{-}5)$$

Assuming that the variation of the volumetric efficiency η_v in Eq, (2-4) is represented by Eq. (2-6), the final equation in the manifold system can be expressed by:

$$\frac{\Delta \eta_v}{\eta_{vo}} = \tau_\eta \frac{\Delta \dot{N}}{N_o} = S\tau_\eta \frac{\Delta N}{N_o} \qquad (2\text{-}6)$$

$$\frac{\Delta \dot{G}_a}{\dot{G}_{ao}} - (1 + S\tau_\eta)\frac{\Delta N}{N_o} = (1 + S\tau_a)\frac{\Delta P_b}{P_{bo}} \qquad (2\text{-}7)$$

Writing Eq. (2-7) in the form of a block diagram, we obtain the portion enclosed by the broken line. Thus, the transfer function $G_n(S)$ is given by Eq. (2-8).

$$G_n(S) = \frac{1}{1 + S\tau_a} \qquad (2\text{-}8)$$

The second term in the left-hand side of Eq. (2-7) means the feedback effect carried out mechanically. In other words, it represents the extent to which the engine speed is restored upon deviation from an equilibrium speed. The manifold absolute pressure (MAP) would increase when the engine speed decreases from an equilibrium speed. As a result, the engine torque would increase and, hence, the speed would increase to the original speed again. Apparently, this is a feedback effect. Although this is one of the most important characteristics of engine, the restoration ability is insufficient as will be shown later.

2.1.2. The transfer function $G_c(S)$ of the fuel control

It is assumed that fuel injection is carried out synchronously with the engine speed. As is well-known, two injection systems are practically used. One is the L-Jetronic system (L-system), in which

the injected fuel quantity is proportional to the air-flow rate per
unit engine revolution. In the other system called D-Jetronic (D-
system), the injected fuel quantity is made to be proportional to
the MAP. The equations for these systems are:

$$P_w = k\frac{\dot{G}_a}{N} \quad \text{(L-system)}$$

(2-9)

$$P_w = kP_b \quad \text{(D-system)}$$

(2-10)

For simplicity, we consider mainly the D-system. Thus, the variation of the injection pulse width is proportional to the MAP and
$G_c(S) = 1$ in principle. However, considering both the stroke lag
and the modulation of the fuel quantity proposed by Hasegawa, we
obtain the following transfer function:

$$G_c(S) = (1 + M(S))\, e^{-S\tau_q}$$

(2-11)

where $M(S)$ is a modulation function; in the case of modulating the
pulse width with the derivative of the manifold pressure, $M(S)$ is
written as follows:

$$M(S) = SK_d T_{inj}$$

(2-12)

In the L-system,

$$M(S) = \frac{S(\tau_a - \tau_\eta)}{1 + S\tau_\eta}$$

(2-13)

2.1.3. The transfer function of $G_f(S)$

The injected fuel flows through the manifold into the engine. Therefore, the trannsport characteristics affect the transfer function, even
with the multipoint injection (MPI), though the effect is slight. On
the other hand, in the single-point injection (SPI), the effect on
$G_f(S)$ is great. As the transport characteristics of the fuel in SPI
system will be formulated later, we consider only the MPI system
here. As shown in Figure 19 the function $G_f(S)$ has two inputs and
single output. This means that it must be written in the matrix form.
However, since our purpose is to derive the transfer function between the air-flow rate and the engine speed, the matrix form is not
used. Assuming that the transport characteristics are given by the

first order lag and lead, the relation among the air-fuel ratio, the air-flow rate into the engine $\dot{G}_e$, and the fuel-flow rate is represented by

$$\frac{\Delta\lambda}{\lambda_o} = \frac{\Delta\dot{G}_e}{\dot{G}_{eo}} - \frac{1 + SK_v\tau_m}{1 + S\tau_m}\frac{\Delta\dot{G}_f}{\dot{G}_{fo}} \qquad (2\text{-}14)$$

where, the first term is the air-flow rate into the engine and the second is the fuel-flow rate into the engine. Rewriting Eq. (2-14), we obtain the following equation.

$$\frac{\Delta\lambda}{\lambda_o} = -S\tau_a\frac{\Delta P_b}{P_{bo}} - \frac{1 + SK_v\tau_m}{1 + S\tau_m}\left(\frac{\Delta P_w}{P_{wo}} + \frac{\Delta N}{N_o}\right) \qquad (2\text{-}15)$$

2.1.4. The engine characteristics ($G_b(S)$)

As with $G_f(S)$, the function $G_b(S)$ is also a three-input and one-output system. Strictly, it also has to be a matrix transfer function. Since the variation of the brake torque ΔTb is given by the linear sum of the torque variation through that of the MAP, the air-fuel ratio, and the engine speed, it is expressed by the following equation.

$$\Delta T_b = \Delta T_{bp} + \Delta T_{b\lambda} + \Delta T_{bn} \qquad (2\text{-}16)$$

$$\Delta T_b = \left\{ K_p\left(\frac{\Delta P_b}{P_{bo}} + \frac{\Delta\eta_v}{\eta_{vo}}\right) + K_\lambda\frac{\Delta\lambda}{\lambda_o}\right\}e^{-S\tau_q}$$

$$+ K_n\frac{\Delta N}{N_o} \qquad (2\text{-}17)$$

where three coefficients K_p, K_λ, and K_n are the three partial derivatives with respect to the MAP, the air-fuel ratio, and the engine speed, respectively.

2.1.5. The transfer function of the flywheel ($G_d(S)$)

The torque variation produces the variation of the engine speed. It is obvious that the characteristics from the torque to the angular velocity are given by Euler's equation:

$$\frac{\pi J}{30}\frac{d\Delta N}{dt} = \Delta T_b \qquad (2\text{-}18)$$

2.1.6. The transfer function $G_e(S)$ from $\Delta\dot{G}_a$ to ΔN

The dead times can be neglected, because they only make the function complex with fruitless results. It is necessary to consider only the essence of the phenomenon of the idle. Thus, neglecting all dead times, the following equations are derived.

$$\frac{\Delta N}{\Delta\dot{G}_a} = \frac{N_o}{\dot{G}_{ao}} \frac{K_p}{K_p - K_n} \frac{1}{1 + bS + cS^2} \tag{2-19}$$

$$b = \frac{\pi N_o J/30 + (K_\lambda \tau_\eta - K_n \tau_a)}{K_p - K_n} \tag{2-20}$$

$$c = \frac{\tau_a\{\pi N_o J/30 - (K_p + K_\lambda)\,\tau_\eta\}}{K_p - K_n} \tag{2-21}$$

It should be noted that the above equation reduces finally to the simple second-order lag. However, as shown in Figure 19, there are some components with the multiple inputs and a output. Moreover, their outputs are given by the linear coupling of their inputs. From the point of the control theory, this is one of the most important properties, and it makes us difficult to understand the phenomena in the idle state. For example, suppose that some phase compensation elements are inserted in the block diagram. If the block diagram is composed of only the components with the single input and single output, the elements operate as expected. That is, the open-loop transfer function is given by the product of the all components. However, if multiple input components are included in the diagram, the elements do not necessarilly operate as expected, because the open-loop transfer function is not given by the product of all components. Such a situation actually occurs in the idle state.

3. Analysis of the ISC system

In this section, we discuss the validity of Eq. (2-19) in comparison with the experimental results on the idle model, propose a practical mathematical idle model, we analyze the stability of the ISC system by using the model, and propose some measures for the improvement of the control result.

3.1. Comparison between the experimental results and Eq. (2-19)

The step response determined experimentally is shown in Figure 21. By using the least-squares method, it was proved that the characteristics can be closely approximated by the following equation which is represented by both the dead times and the second-order lag.

$$\frac{\Delta N}{\Delta \dot{G}_a} = \frac{k \cdot e^{-SL}}{1 + 2\zeta S/\omega_n + (S/\omega_n)^2} \qquad (2\text{-}22)$$

This result shows a good agreement with that of Takahashi except for the term of the dead time. It also agrees with Eq. (2-19) very well. Moreover, a good agreement is obtained between the natural angular frequency in Eq. (2-19) and the measured one. Therefore, it is considered that the rational part of the transfer function in the idle is given by Eq. (2-19) and that the total transfer function is represented by Eq. (2-22)

3.2. Analysis of the stability of ISC system

Now let us analyze the stability of ISC system. For this purpose, a simplified block diagram of the ISC system is given in Figure 22. The open-loop transfer function $G_o(S)$ is

$$G_o(S) = G(S)\, G_a(S) G_e(S)$$

$$= \frac{1 + SK_cT_i}{ST_i}\, e^{-SL} \frac{k}{(1 + ST)^2} \qquad (2\text{-}23)$$

where for simplicity we assume $G_a(S) = 1$. Also the rational part of the transfer function $G_e(S)$ is given by $k/(1 + ST)^2$, because the resistance coefficient ζ is approximately equal to unity by the experimental result. Therefore, the characteristic equation is given by the following equation.

$$1 + e^{-SL} \frac{1 + SK_cT_i}{ST_i} \frac{1}{(1 + ST)^2} = 0 \qquad (2\text{-}24)$$

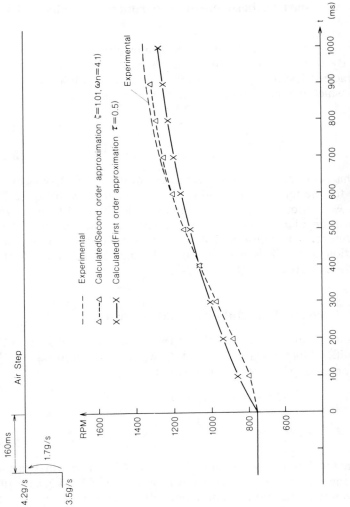

Figure 21. Step response of idle speed against the stepwise variation of the air-flow rate

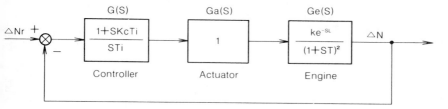

Figure 22. A simplified block diagram of ISC system

where the value of k is assumed to be unity, because the value is negligible for our purpose. Subsituting $S = j\omega$, and representing the second term by $G_t(j\omega)$, we obtain:

$$G_t(j\omega) = e^{-j\omega T \cdot L_n} \times \frac{1 + j\omega T \cdot K_c T_n}{j\omega T \cdot T_n(1 + j\omega T)^2} \qquad (2\text{-}25)$$

where

$$L_n = L/T, \; T_n = T_i/T,$$

i.e., the normalized dead time and the normalized integral time of the controller, respectively. The stability of the ISC system can be analyzed by drawing the Nyquist diagram of the function $G_t(j\omega)$. When we assume $T = 0.3$ sec and $L = 0.16$ ($N_o = 750$ rpm) to be a typical condition, L_n is nearly equal to 0.5. Figure 23 shows the Nyquist diagram in case of the I-control ($K_c = 0$), which points out the following:

(1) In case of $T_n = 0.5$ (the broken line), the ISC system causes the hunting with a frequency of about 0.4 Hz, which agrees well with the observed value.
(2) In case of $T_n = 1.0$ (the solid line), the system is in the stable limit, because the line runs through the vicinity of the point $(-1,0)$. Therefore, in this system it is impossible to increase the integral gain over $T_n = 1$ ($T_i = T$)
(3) The the solid line and the broken line cross the real axis at one point, i.e., at the same frequencies; both have the same phases at one frequency, which is one of the characteristics of I-control as shown by the next equation.

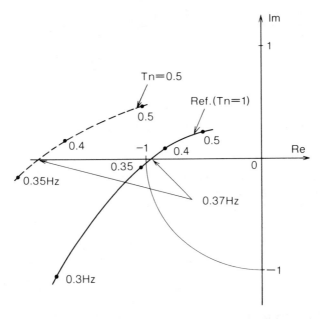

Figure 23. Examples of Nyquist diagram of ISC with I-control ($L_n=0.5$ and $T=0.3$ Sec; second order lag)

$$\arg G_t(j\omega) = -\omega T L_n - \tan^{-1}\frac{2\omega T}{1 - (\omega T)^2} \qquad (2\text{-}26)$$

(4) The phase lag due to the second order lag is about seven times as large as that due to the dead time. Calculating the phase lags at 0.3 Hz we have the following equations.

$$TL_n = 0.37 \times 2 \times 0.3 \times 0.5 = 0.35\text{rad}$$

$$\tan^{-1}\frac{2 \times 2.325}{1 - 2.325^2} = 1.22\text{rad}$$

We consider next the Nyquist diagram of the PI-control. The Nyquist diagrams against the different proportional gains K_c at the normalized integral time Tn of unity is shown in Figure 24; the following points are easily understood.

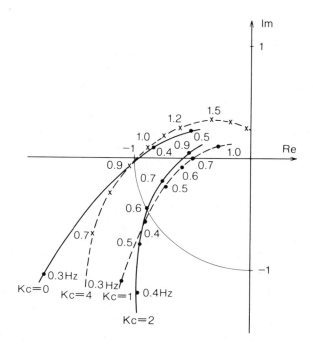

Figure 24. Examples of Nyquist diagram of ISC with PI-control ($L_n=0.5$ and $T_n=1$; second order lag)

(1) The lines for $K_c = 1$ and 2 show the most stable system.
(2) The larger the gain K_c becomes, the higher become the frequencies of the real axis.
(3) These phenomena can be explained by the competition between the phase lead effect and the increase in the absolute value of the term $1 + SK_cT_i$ in Eq. (2-24).

In other words, the phase lead effect of $1 + SKcTi$ in the controller compensates the phase lag due to the second order lag without increasing the the absolute value of $G_t(j\omega)$ remarkably, when the gain K_c is relatively low. As a result, the frequencies at which the lines cross the real axis can be increased and this brings about the decrease in the total gain by the the term ST_i. Thus, the stability increases compared with I-control. On the other hand, the line for $K_c = 4$ shows that the system becomes unstable again. For larger gains, the effect of increasing the absolute value of $1 + SK_cT_i$ in Eq. (2-24) becomes dominant. Thus, the frequencies at the real axis vary slightly with the increase in the prportional gain K_c, and the total gain becomes large.

(4) The PI-control can also improve the response of the ISC system, because an increase in the gain K_c raises the cross frequency at which the Nyquist diagram crosses the unit circle whose center exists at the origin.

(5) After all, the PI-control can improve both the stability and the response of the ISC system. However, the extent of the improvement is not so remarkable as expected. This is because the phase lag of the controlled object represented by Eq. (2-22) is very large. Other measures are neccessary for remarkable improvement of the performance characteristics of the ISC system.

Then, the question is "what is the satisfactory level for improvement?" To estimate this, we studied the Nyquist diagram of the first-order lag. Figure 25 shows the Nyquist diagram of the first order lag. In this case, the open transfer function is given by Eq. (2-27).

$$G_t\,(j\omega) = e^{-j\omega T^*L}\,{}_n\frac{1 + j\omega T^*K_cT_n}{j\omega T^*T_n(1 + j\omega T^*)} \qquad (2-27)$$

According to Figure 25, the performance characteristics are greatly improved. For example, (1) the frequency at which the line crosses the real axis is 1.7Hz and the gain margin is 20dB, (2) the cross frequency is about 0.6Hz. In comparison with the case of $K_c = 1$ in Figure 24 where the gain margin is 6dB and the cross frequency about 0.4Hz, it can be seen that both the stability and the response are greatly improved.

4. A proposal of a local feedback compensation

As mentioned above, the control results of the conventional ISC system are determined mostly by the phase lag of the controlled

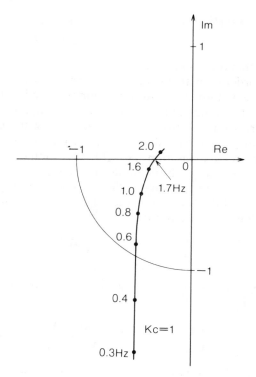

Figure 25. An example of Nyquist diagram of ISC with PI-control (L = 0.5 and T = 1; first order lag)

object and the simple phase compensation is not so effective as to improve control results remarkably; other measures are necessary which can change the properties of the controlled object. We will propose a local feedback about the state variable.

4.1. Formulation of the local feedback compensation

For the consideration of the essence of the second-order lag, we review the formulation of the Section 2.1. In Eq. (2-7) equating $\tau_\eta = 0$, for simplification, we obtain:

$$\frac{\Delta P_b}{P_{bo}} = \frac{1}{1 + S\tau_a} \frac{\Delta \dot{G}_a}{\dot{G}_{ao}} - \frac{1}{1 + S\tau_a} \frac{\Delta N}{N_o} \qquad (2\text{-}28)$$

This shows that the manifold absolute pressure MAP is represented by the first order lag with the time constant τ_a. This means physically that the air flows into the engine after being charged in the manifold volume V_m. This is one phase lag existing in the idle state. Another phase lag involved is given by the Euler equation which converts the torque to the engine speed. Apparently, this process cannot be changed by any compensation. Therefore, we should try to compensate the phase lag due to the manifold effectively.

Suppose that we superimpose the air-flow rate proportional to the derivative of the MAP upon the original air flow rate $\Delta \dot{G}_{ap}$. We obtain the following equation:

$$\frac{\Delta \dot{G}_a}{\dot{G}_{ao}} = \frac{\Delta \dot{G}_{ap}}{\dot{G}_{ao}} + S\tau_a \frac{\Delta P_b}{P_{bo}} \qquad (2\text{-}29)$$

Substituting this equation into Eq. (2-7), the following equation can be obtained.

$$\frac{\Delta \dot{G}_{ap}}{\dot{G}_{ao}} - (1 + S\tau_\eta) \frac{\Delta N}{N_o} = \frac{\Delta P_b}{P_{bo}} \qquad (2\text{-}30)$$

According to this equation, we can easily see that the first-order lag between the air-flow rate and the MAP is eliminated effectively. Therefore, the transfer function of the idle is expected to be reduced to the first-order lag. Let us confirm this expectation. Deriving the transfer function of the idle in the same manner as shown in deriving Eq. (2-19), we finally obtain the following equation.

$$\frac{\Delta N}{\Delta \dot{G}_{ap}} = \frac{N_o}{\dot{G}_{ao}} \frac{K_p}{K_p - K_n} \frac{1}{1 + S\tau^*} \qquad (2\text{-}31)$$

$$\tau^* = \pi N_o J / \{30(K_p - K_n)\} \qquad (2\text{-}32)$$

It should be noted that the expectation described above is satisfied fully.

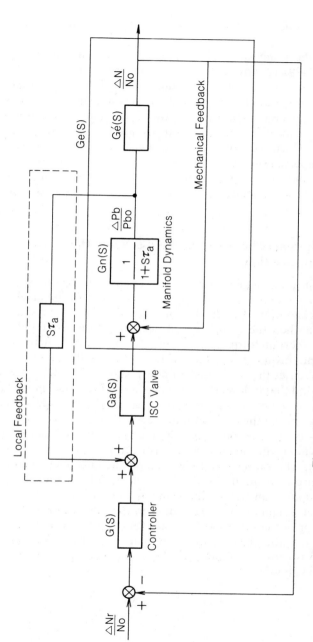

Figure 26. A block diagram ISC with local feedback

Thus, the remarkable effectiveness of the compensation given by Eq. (2-29) is shown. Now, this compensation is a local feedback, because we can readily obtain the block diagram of the ISC system compensated by Eq. (2-29). This compensation is shown in Figure 26 by a rectangle enclosed with broken lines. Apparently, it is a kind of local feedback of state variables such as the manifold pressure. Thus, we can define this compensation as a local feedback compensation. We study in the next section both its physical meaning and the differences between the feedback compensation and the phase compensation widely used in the design of the controller.

4.2. Differences between the feedback and the phase compensation

The physical meaning of feedback compensation becomes clear by considering its operation; it operates so as to assist the mechanical feedback mentioned in 2.1.1. For further detail, let's consider an example: when the engine speed decreases from the equilibrium point by a certain torque disturbance, MAP naturally increases and the engine torque also increases. It will be readily seen that when MAP increases the additional air flow given by Eq. (2-29) becomes positive and that it flows into the manifold. This results in the faster recovery of the manifold pressure. Also, the engine speed can recover faster than the case without the additional air flow. Thus, the feedback compensation assists the mechanical feedback. In other words, it alters the engine characteristics by assisting the mechanical feedback. Therefore, it can always alter the engine characteristics and reduce it to the first-order lag. On the other hand, the phase compensation can reduce the characteristics to the first order lag only when a same factor exists in both the numerator and the denominator of the open-loop transfer function $G_t(S)$. For example, suppose that the phase compensation of $1+ST_c$ is carried out. This situation is shown in Figure 27. Then $G_t(S)$ is represented by the equation below.

$$G_t(S) = \frac{1 + SK_cT_i}{ST_i}(1 + ST_c)\frac{ke^{-SL}}{(1 + ST)^2} \qquad (2\text{-}33)$$

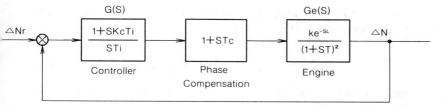

Figure 27. A block diagram ISC system with conventional phase compensation

Apparently, only when $T_c = T$ the factor $1 + ST_c$ can be cancelled. This is one of the greatest differences between the feedback and the phase compensation.

4.3. Generalization of feedback compensation and the effect of dead time on the reduced transfer function

As shown in the proceeding section, feedback compensation is very effective. Is it always effective under any condition? In order to answer this question, a generalization of this method is given below. From Eq. (2-7) the following equation holds.

$$\Delta P_b^* = \frac{1}{1 + S\tau_a} \Delta \dot{G}_a^* - \frac{1 + S\tau_\eta}{1 + S\tau_a} \Delta N^* \qquad (2\text{-}34)$$

where, the asterisk denotes the normalized variations, for example:

$$\Delta P_b^* = \frac{\Delta P_b}{P_{bo}}$$

Writing the feedback compensation function as $C(S)$, we obtain the block diagram shown in Figure 28. In the figure, the transfer function from ΔP_b^* to ΔN^* is represented by $G_e'(S)$, and the transfer function $G_m(S)$ shows the mechanical feedback due to the mass conservation law in the manifold. By using the block diagram we obtain the transfer function between $\Delta \dot{G}_{ap}^*$ and ΔN^* as follows.

$$G_e(S) = \frac{\Delta N^*}{\Delta \dot{G}_{ap}^*} = \frac{1}{G_m(S) + \{1 + S\tau_a - C(S)\}/G_e'(S)} \qquad (2\text{-}35)$$

Thus, we readily understand that: (1) if $C(S) = 0$, the transfer function is given by Eq. (2-36)

$$G_e(S) = \cfrac{1}{G_m(S) + (1 + S\tau_a)/G'_e(S)} \qquad (2\text{-}36)$$

and (2) if $C(S) = S\tau_a$ the transfer function can be expressed by the following equation

$$G_e(S) = \cfrac{1}{G_m(S) + 1/G'_e(S)} \qquad (2\text{-}37)$$

The function $G'_e(S)$ is derived by Eqs. (2-17) and (2-18);

$$\Delta P_b^* = G'_e(S)^{-1} \Delta N^*$$

$$= \cfrac{SN_o J\pi/30 - K_n}{K_p} \Delta N^* \qquad (2\text{-}38)$$

where dead time τ_q is neglected. Therefore, in case (1), the transfer function of engine is represented by the second-order lag and in case (2) by the first-order lag. This means that the order of Ge(S) becomes equal to that of either $G'_e(S)^{-1}$ or $G_m(S)$ by feedback compensation. In other words, from Eq. (2-38) and Figure 28, we can always reduce the engine characteristics to the first-order lag by feedback compensation. Therefore, it is understandable that feedback compensation is useful for general purposes.

At the end of this section, we discuss an interesting matter shown by Eq. (2-38). As can be seen, the transfer function $G'_e(S)$ between ΔP_b^* and ΔN^* includes the negative coefficient of $-K_n$, which originates from Eqs. (2-17) and (2-18). By equating these equations, we obtain the relation. Therefore, physically the characteristics of the multiple inputs and single output given by the linear combination of the inputs produce the negative sign. On the other hand, according to stability control theory, the negative coefficient means that the system is unstable.

Although $G'_e(S)$ has a negative sign, the engine is stable to some extent. This is attributable to the effect of the mechanical feedback. The control theory teaches that the feedback is useful generally to the stabilization of a system. This is also true with the operation of engine.

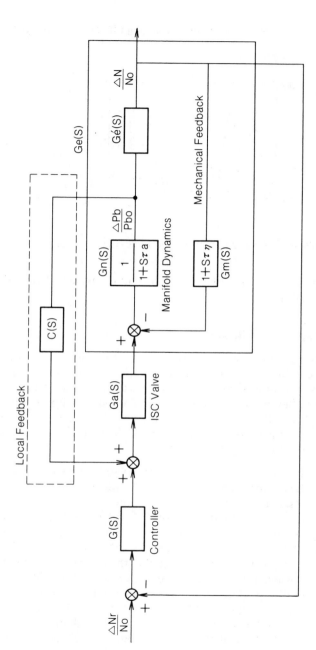

Figure 28. Generalization of local feedback

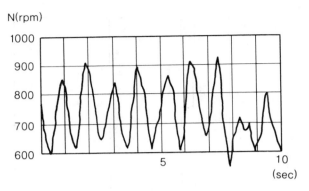

Figure 29. An example of the periodic oscillation of idle speed

5. Application of the model to the analysis of the idle stability

5.1. Aspects of idle stability

Starting with the aspects of idle stability, we often meet the periodic oscillation of the idle speed as shown in Figure 29, especially in D system (speed density system) with a manual transmission. But this phenomenon never occurs in a system with an automatic transmission. It is reasonable to imagine that the periodic oscillation is not considered serious in the U.S.A. The periodic oscillation has the following characteristics:

(1) The frequency ranges from 0.1Hz to about 1.0Hz.
(2) The larger the manifold volume, the less the frequency and the larger the amplitude of the oscillation.
(3) In the multiple point injection (MPI), the amplitude is weakly dependent on the air-fuel ratio, although it increases both at the lean and the rich end. On the other hand, in single point injection (SPI), it is rather small at the rich side. But the leaner the air-fuel ratio, the larger the amplitude as shown in Chapter 3.
(4) Generally speaking, the hunting phenomenon hardly occurs in the engine equipped with an automatic transmission.
(5) Without the ISC system the periodic oscillation of the idle speed occurs. The periodic oscillation is obviously different

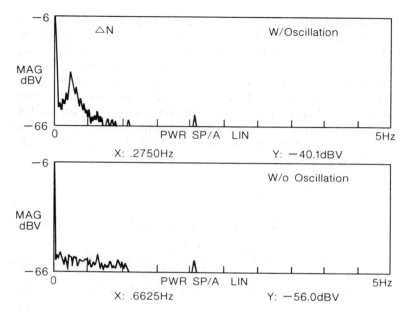

Figure 30. Examples of frequency spectra of the variation in idle speed

from the hunting caused by an unstable ISC system with high control gain as explained in the preceding sections.

(6) It cannot be removed by the present ISC system.

Figure 30 shows the frequency spectra with and without oscillation. It should be noted that there is the large frequency component with about 0.3Hz. This means that the frequency component would be excited by certain causes. In the next section, by using the idle model which provides promising control results for the ISC system, the analysis and the measure for the reduction of this phenomenon are presented.

5.2. Formulation and negative gain margin as an index of periodic oscillation

Periodic oscillation occurs independently of the ISC system as mentioned above; it occurs essentially without variation of the air-flow rate through the throttle or the bypass air passage. This suggests that the variation of the air-flow rate can be equal to zero in the analysis of the idle stability. Moreover, it must be equal to zero, because the flow in the throttle is choked and the opening is not changed. This point is one of the most different points from the ISC system. Keeping only this in mind, the model for the explanation of this phenomenon is essentially the same as that for the ISC system. Therefore, we rewrite the pertinent equations involved in the model by taking into account the dead time neglected in the preceding section.

$$\Delta P_b^* = -\frac{1 + S\tau_\eta}{1 + S\tau_a} \Delta N^* \tag{2-39}$$

$$\Delta P_w^* = (1 + M(S))e^{-S\tau_c}\Delta P_b^* \tag{2-11}$$

$$\Delta\lambda^* = -S\tau_a\Delta P_b^* - \frac{1 + SK_v\tau_m}{1 + S\tau_m}(\Delta P_w^* + \Delta N^*) \tag{2-15}$$

$$\Delta T_b = \Delta T_{bp} + \Delta T_{b\lambda} + \Delta T_{bn} = \{K_p\Delta(\eta_v P_b)^* + K_\lambda\Delta\lambda^*\}e^{-s\tau_q} + K_n\Delta N^* \tag{2-17}$$

$$SJ\Delta N^* = 30/(\pi N_o)\Delta T_b \tag{2-18}$$

The block diagram obtained is shown in Figure 31, which has two different points compared with the ISC system. It has no controller like PI controller. Another point is that the variation of the engine speed is directly connected to that of the manifold absolute pressure MAP, because there is no variation of the air-flow rate through the throttle valve. Thus, in this case, the complete closed loop is constructed as shown in Figure 31.

For solving the oscillation problem which occurs in such a closed loop, we consider two different ways. One is forced oscillation, and the other is self-exciting. Forced oscillation is the model where the system oscillates by the applied disturbance. Therefore, in this model, the less the disturbance the less the oscillation amplitude. In the actual engines, various torque variations can be considered as disturbances: for example, the cyclic variation and various electric loads. On the other hand, self-exciting oscillation is the model where the system will oscillate without any disturbance. In actual

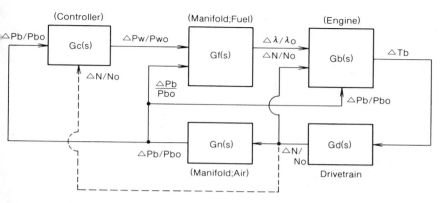

Figure 31. A linearized model of the idle state for explaining idle stability

engines, there are many disturbances that will excite both oscillations. Thus, it is meaningless to discriminate between forced and self-exciting oscillations. One of the most important matters is that we can get some indices concerning the stability of the system shown in Figure 31 and that by the indices we can explain the phenomenon and find the countermeasure. Seeking the indices of stability from this point of view, we easily think of the gain margin that is widely used in the analysis of the control system. As is well-known, if the gain margin is negative the system will oscillate spontaneously, and if it is positive the system will not oscillate. However, the larger the gain margin, the smaller the variation against the disturbance. Therefore, this is a suitable index for the analysis of the idle stability.

In the present discussion, we try to explain the aspects of the idle stability by using the loop gain that is defined by the negative gain margin or the next equation referring to Figure 32.

$$\text{Loop gain} = \Delta G_o = -g_m = -\log OC = 20 \log G_t (j\omega_o) \quad (2\text{-}40)$$

$$\arg G_t(j\omega_o) = 2n\pi \quad (2\text{-}41)$$

The system will oscillate by the self-exciting oscillation when the loop gain is positive, and when the loop gain is negative the system oscillates not by self-exciting oscillation but by forced oscillation.

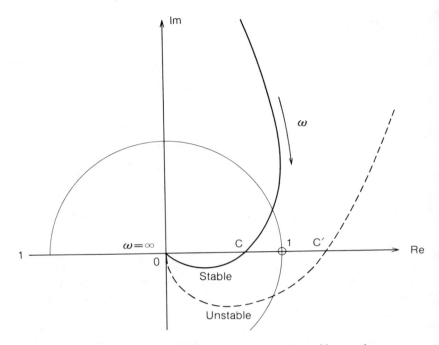

Figure 32. Nyquist diagram for the definition of loop gain

The loop gain is defined as the absolute value of the open-loop transfer function $G_t(j\omega)$ at a frequency that satisfies the second equation (2-41). It is often called the phase condition and shows the frequency ω_o at which the total phase difference becomes n times as much as 2π. This means that, when some variation is applied to some place in the block diagram shown in Figure 31, the phase difference between the variation induced there by the original variation and the original one reduces to zero. For example, let us suppose that we open the loop in Figure 31 between the output of the drive train and the input into the manifold system, and we apply some disturbance ΔN^* to the input. The induced variation at the output $\Delta N'^*$ has the same phase as the original disturbance ΔN^* at the frequency ω_O predicted by the phase condition (Eq. (2-

41)). Therefore, if the loop gain is positive, the induced variation at the output $\Delta N'^*$ becomes larger than the original variation ΔN^*, and thus the system oscillates. Apparently, the larger the loop gain, the larger the oscillation amplitude. Even if the loop gain is negative, the larger the loop gain, the larger the influence of the disturbance and the total amplitude increases with increasing the loop gain.

The analysis method with the loop gain can be applied to analysis of the oscillation both in the linear and the nonlinear system.[35] Further, since it is carried out in the frequency region, we can easily obtain the physical image by drawing Eqs. (2-17) and (2-18) on the complex plane.

We now analyze the phenomenon. The open-loop transfer function $G_t(j\omega)$ is obtained by Eqs. (2-17) to (2-18).

$$G_t(j\omega) = \frac{30/\pi}{SN_oJ}(\Delta T_{bp} + \Delta T_{b\lambda} + \Delta T_{bn})$$ (2-42)

$$\Delta T_{bp} = -\frac{K_p(1 + \omega^2\tau_a\tau_\eta)}{1 + S\tau_a}\Delta N^*$$

We can readily understand whether the system is stable or not by drawing the Bode diagram. For this purpose, it is necessary either to identify or determine the parameter values included in the open-loop transfer function. The values obtained are listed in Table 11. These values were determined by the experiments and the theoretical calculation as mentioned in the reference[35]-[38].

An example of the calculated Bode diagram is shown in Figure 33. The broken line shows the phase of $G_t(j\omega)$ against frequencies. Whereas the solid line in Figure 33 shows the loop gain plotted also against frequencies. From this figure the following points are revealed:

(1) The value of the frequency $\omega_o = 2\pi f_o$ is 0.15Hz and the loop gain at the frequency has the positive value of 7.5dB.

(2) Therefore, the system oscillates spontaneously and the engine speed oscillates as in Figure 29 with a frequency near 0.15Hz, which is close to the observed value of 0.2Hz.

(3) In this case, idle stability is determined by self-oscillation. A large oscillation of the idle speed occurs.

Table 11. Values of
parameters (Two-cylinder
engine).

Parameter	Value
K_n	1.0 (Kgm)
K_p	1.75 (Kgm)
K_λ	-1.45 (Kgm)
V_m	1.2 (litre)
V_h	0.55 (litre)
τ_a	0.58 (Sec)
τ_m	0.5 (Sec)
τ_c	0.055 (Sec)
τ_q	0.060 (Sec)
η_{vo}	0.60
τ_η	0.12 (Sec)
N_o	750 (RPM)
P_{bo}	360 (mmHg, abs)
λ_o	1.0
J	0.0056 (KgmS^2)

As mentioned above, the value of the loop gain represents the oscillation amplitude. Therefore, seeing the value of the loop gain we can easily understand the dependency of the idle stability on the various parameters.

For example, Figure 34 shows the loop gain with varying manifold volume; the larger the manifold volume the larger the loop gain. This explains the fact that the larger the manifold volume, the larger the amplitude of the engine speed ΔN_{pp}. In another example, there is the dependency of the idle stability on the coefficient of the derivative fuel modulation proposed by Hasegawa.[39] Since in the derivative fuel modulation the injection pulse width is modulated proportionally to the derivative of the manifold pressure, the modulation function is given by

$$M(S) = SK_dT_{inj}$$

where K_d is a coefficient of the derivative fuel modulation.

Considering the modulation, the loop gain against K_d is represented by Figure 35. The broken line shows the experimental re-

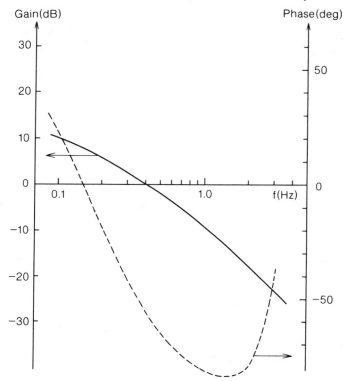

Figure 33. An example of Bode diagram

sults and the solid line the predicated value of the loop gain. Both lines show that an optimum value of K_d existing near the value 6, and we see a good agreement between the experimental and the predicted values. In this figure the loop gain over 3 is negative. This means that the system generates no self-exciting oscillation in that region of K_d and the idle stability is determined by the torque disturbance mentioned above. In other words, in this region the forced oscillation is generated in the system.

In this manner, the aspects of the idle stability mentioned in the Section 5.1, especially the first two items were explained. Since it is

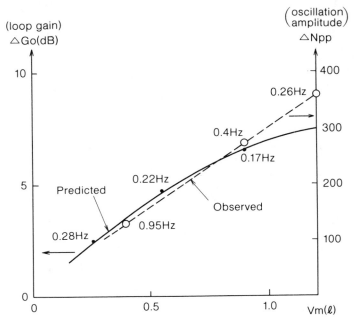

Figure 34. Comparison between the calculated loop gain (ΔG_o) and the observed oscillation amplitude (ΔN_{pp})

shown later that another aspect can be also explained, the physical image of idle stability is explained in the next section.

6. The physical image of idle stability

6.1. Proposal of a vector diagram

The Bode diagram is very useful to derive the loop gain quantitatively and it also gives us a physical image of generating the oscillation, when every component in the system is composed of a single input and a single output, because both the loop gain and the phase of the open loop transfer function are given by the summations of the gain and the phase of every component. However, the idle

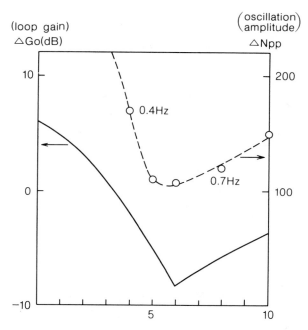

Figure 35. Both behaviors of the loop gain (ΔG_o) and the observed oscillation amplitude (ΔN_{pp}) against K_d

system includes several components with multiple inputs and single output given by the linear summation of their inputs. Therefore, the open-loop transfer function is not necessarily given by the product of every component.

As is well-known, the summation can be shown graphically in the complex plane; if we draw the vector locus of every component in the complex plane, the physical image of the idle system becomes clear. A typical example of the vector loci in the complex plane is drawn in Figure 36. These vector loci are obtained by the following equations:

$$\Delta T_b = \Delta T_{bp} + \Delta T_{b\lambda} + \Delta T_{bn}$$

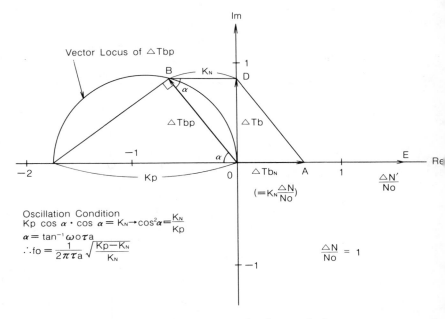

Figure 36. An example of vector loci

$$\Delta T_b = \Delta T_{bp} + {}_\Delta T_{b\lambda} + \Delta T_{bn} \tag{2-43}$$

$$\Delta T_{bp} = - \frac{K_p}{1 + S\tau_a} \Delta N^* \quad (\text{for } \tau_q = \tau_\eta = 0) \tag{2-44}$$

$$\Delta T_{b\lambda} = 0 \quad (\text{for } \tau_q = \tau_\eta = 0) \tag{2-45}$$

$$\Delta T_{bn} = K_n \Delta N^* \tag{2-46}$$

$$\Delta N^* = (30/\pi)\Delta T_b/SN_o J \tag{2-47}$$

where all dead times due to the stroke are assumed to be zero.

The torque variation ΔT_{bn} is induced by the variation of the engine speed ΔN^*; for instance it is generated by the torque increase with increasing idle speed. The torque increase occurs by the negative characteristics of the friction of the engine or the positive characteristics of the indicated mean effective pressure against the engine revolution.

The torque variation ΔT_{bp} results from the variation of the charged air into the engine. It is represented the variation of $\Delta(\eta_v P_b)$, and thus, easily given by Eq. (2-44) from Eqs. (2-6) and (2-8) under $\tau_\eta = 0$. Generally, the vector locus of the first order lag is given by a semicircle. This property is expressed by the the semicircle in the second quadrant in Figure 36. From the vector locus of ΔT_{bp}, we can obtain easily the vector for some frequency by connecting with the straight line between the origin and the point B for the frequency on the locus. The vector **OB** thus obtained gives the torque variation induced by the variation of the engine speed.

In the case considered here, the torque variation $\Delta T_{b\lambda}$ is reduced to zero because both the dead time and the value are assumed to be zero and the fuel injection is carried out proportionally to the manifold absolute pressure (speed-density).

Equation (2-43) shows that the variation of the brake torque ΔT_b is given by the summation of ΔT_{bn}, ΔT_{bp}, and ΔT_b, and is easily expressed by the parallelogram law in the complex plane; the variation of the brake torque is given by the vector ΔT_b in Figure 36.

Now, the variation of the engine speed induced the torque variation ΔT_b is represented by Eq. (2-47), which shows that we can obtain the speed variation ΔN^* both by multiplying the magnitude of the ΔT_b by a factor of $30/\pi N_o J$ and rotating the vector ΔT_b by 90 degrees toward the real positive axis (CW). Therefore, the vector $\Delta N'^*$ in Figure 36 represents the variation of the engine speed. Then the angle between the vector $\Delta N'^*$ and the real axis expresses the phase difference between $\Delta N'^*$ and ΔN^*, which is the phase of the open-loop transfer function $G_t(j\omega)$. Therefore, if the vector $\Delta N'^*$ agrees with the real axis, the phase difference becomes zero and the phase condition is satisfied. Then, the division of $\Delta N'^*$ to ΔN^* gives the loop gain defined by Eq. (2-40). In other words, the vector diagram drawn by the above processes is completely equivalent to the Bode diagram.

From Figure 36 we can understand the following points:

(1) The dominant cause of the hunting phenomenon of the idle speed are both the torque variation ΔT_{bp} generated by the phase lag due to the manifold volume and the torque variation ΔT_{bn} due to the variation of the engine speed.

(2) The larger the manifold volume, the larger the loop gain, as

mentioned above. In other words, when the manifold volume increases, the frequency at the point B decreases, because the time constant τ_a increases as can be seen from Eq. (2-5). The decrease of the frequency results in the increase of the factor $30/\pi S N_o J$, and thus results in the the increase of the loop gain.

(3) The larger the parameter K_n, the larger the loop gain. On the other hand, the larger K_p, the the smaller the loop gain. That is, when K_n increases the point B moves toward the left on the semicircle and the frequency of the point of B that satisfies the phase condition Eq. (2-41) decreases. Therefore, the loop gain increases as in the case in which the manifold volume was increased above. On the contrary, when K_p increases, the loop gain decreases; then the semicircle becomes large and the frequency at the point of B increases in Figure 36. This decreases the factor of $30/\pi S N_o J$, and decreases the loop gain.

(4) The reduction effect of the derivative fuel modulation on the hunting phenomenon is also explained intuitively by the vector diagram. We will explain this in detail in the next section. As mentioned above, by the vector diagram several dependencies of the idle stability on several parameters are explained with great ease, simplicity, and intuitiveness. Therefore, this technique is also useful to improve the idle stability.

6.2. Explanation of the effect of the derivative fuel modulation

As shown in Figure 35, derivative fuel modulation is very effective in reducing the loop gain. In this section, we show the effectiveness of derivative fuel modulation using the vector diagram of a simplest case. Here, we assume that both the all dead times and the value of τ_η are equal to zero. One of the most different points between cases with and without derivative fuel modulation is the modulation results in the torque variation $\Delta T_{b\lambda}$ which reduces the loop gain. This important point is shown below.

The torque variation $\Delta T_{b\lambda}$ is given by

$$\Delta T_{b\lambda} = K_\lambda \frac{S K_d T_{inj}}{1 + S\tau_a} \Delta N^* \tag{2-48}$$

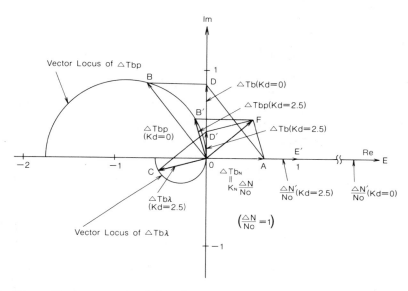

Figure 37. An example of the effect of the derivative fuel modulation on the idle stability by a vector diagram

The vector locus is drawn as a semicircle in the fourth quadrant as shown in Figure 37. Apparently, the vector ΔT_b is reduced by the vector $\Delta T_{b\lambda}$ and the frequency ω_o which satisfies the phase condition increases as shown by the shift from the point B to B'. This reduces the loop gain in the same manner as mentioned above.

In Figure 37, the larger the coefficients of the derivative fuel modulation K_d becomes the smaller the loop gain becomes. Therefore, finally, the phase condition does not become satisfied. Actually this does not occur, since the dead time exists. If it is assumed that all vectors of the torque variations except the torque variation ΔT_{bn} circulate around the origin, the vector ΔT_{bp} and $\Delta T_{b\lambda}$ come to appear in the first and the second quadrant, respectively. When the dead times such as τ_q are considered, for example, the vector loci of ΔT_{bp} and ΔT_b are shown in Figure 38 and Figure 39, respec-

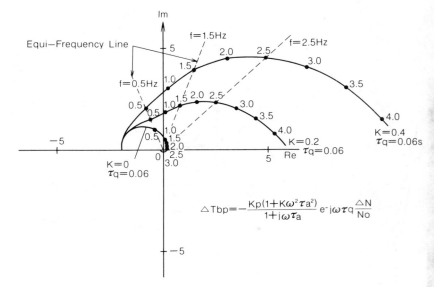

$$\triangle Tbp = -\frac{Kp(1 + K\omega^2 \tau a^2)}{1 + j\omega \tau a} e^{-j\omega \tau q} \frac{\triangle N}{No}$$

Figure 38. Vector loci of ΔT_{bp} with dead time

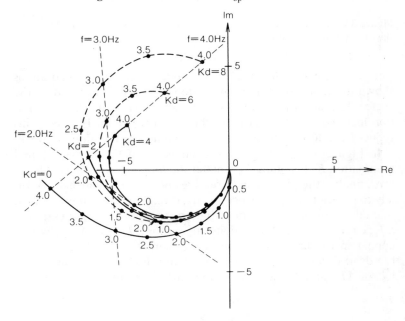

Figure 39. Vector loci of $\Delta T_{b\lambda}$ with dead time

tively. When these vector loci are represented as shown in these figures, the vector $\Delta T_{b\lambda}$ of the torque vaiation due to the air-fuel variation induced by the variation of the engine speed increases the total torque variation ΔT_b. Therefore, the loop gain has a minimum value against the magnitude of the derivative fuel modulation K_d. This is the physical explanation for the optimum value of K_d shown in Figure 35.

7. Several measures to improve idle stability

We have explained the idle stability of the MPI system so far, and not referred to the idle stability of the single point injection (SPI). The idle stability in the SPI system has a close relation to the fuel transportation characteristics in the intake manifold. Moreover, the atomization characteristics of the fuel injector has a great influence on the idle stability. Therefore, both the idle stability in the SPI system and its improvement will be discussed in detail in Chapter 3.

Now, several means to improve the idle stability of the MPI and the SPI systems will be discussed. In Table 12, all means considered are summarized. We point out below the advantages of these measures.

(1) The aim of increasing the operating engine speed is to decrease the time constant τ_a and thereby reduce the loop gain. The increase in the idle speed often increases the fuel comsumption even if the comsumption rate is reduced. This measure is useful to both the MPI and the SPI system.

(2) To operate the engine at richer air-fuel ratios is more effective in the SPI system than the MPI system. As shown later the aim of this measure is to remove the effect of the torque variation due to the variation of the air-fuel ratio. Unfortunately, this measure also increases the fuel comsumption.

(3) Increasing the inertia moment of the flywheel aims to decrease the factor $30/\pi N_o J$ and is effective for both the MPI and the SPI system. However, it increases the rise time of the vehicle speed on acceleration. Except for this disadvantage this method is very effective.

Table 12. Measures for improving idle stability.

	Measures	Advantages or disadvantages
Engine system	1) Increase in engine speed	Often increases fuel consumption Effective to MPI and SPI
	2) Enrichment of A/F	Often increases fuel consumption Effective to SPI
	3) Increase in inertia moment	Often causes slow response Effective to MPI and SPI
	4) Stabilization of combustion	Effective in negative loop gain
Control system	1) Control of air flow rate	High response ISC valve required Effective to MPI
	2) Modulation of injected fuel	Inconsistent with A/F control Often causes oscillation at another operating point of engine Effective to MPI and SPI
	3) Spark timing control	Consistent with A/F control Increases fuel consumption
Induction system	1) Increase in K_v	Effective to SPI Better atomization injector required
	2) Decrease in manifold volume	Effective to MPI Decreases charging efficiency at WOT

(4) The stabilization of the combustion in the cylinder reduces the disturbance of the torque variation. This method is useful in the region of the negative loop gain, because in the region of the positive loop gain the system oscillates spontaneously without the disturbance. It would be one of the most difficult ways to be performed.

(5) The control of the air-flow rate is one of the most effective means for reducing the variation of the idle speed; it aims at the compensation of the phase lag in the torque variation ΔT_{bp}. The principle of this measure is the same as the improvement of the ISC system, i.e., by controlling the air-flow rate according to the derivative of the manifold absolute pressure as shown in previous section. This method has two disadvantages.

One is the requirement of the ISC valve with good frequency characteristics, and the other is the requirement for a wide dynamic range.

(6) The derivative fuel modulation is effective in the region of the positive loop gain. The principle of this method is to generate the torque variation $\Delta T_{b\lambda}$ with an appropriate phase difference. Therefore, the dead times in the system produce some limit in the effect as mentioned above. Moreover, it may be inconsistent with the control of the air-fuel ratio. Also, it may cause other oscillations at another operating point; for example, when the vehicle runs at the lowest speed with the idle throttle open, another oscillation often occurs.

(7) The control of spark timing is one of the most effective methods. It affects the idle stability in the fastest manner compared with other methods. And the effect is remarkable. Moreover, it is consistent with the control of the air-fuel ratio, because this method generates the torque variation keeping the air-fuel ratio constant. The disadvantages of this method are a narrow control range, the need for an electronic spark timing control system and the slight increase in the fuel consumption.

(8) Increase in the air-borne rate K_v of the fuel is reduced by the delay of the fuel transportation. This means, the improvement of the fuel injector in SPI system, and moreover, in MPI system, the improvements of the injection timing, and the direction and the position of the injector. This method is effective in the region of positive loop gain in both injection systems. Technically this method is one of the most difficult ones.

(9) The reduction of the manifold volume aims at the decrease in the time constant τ_a. Thus it prevents the phase lag of the torque variation of ΔT_{bp}. Also, it may reduce the inertia effect in the manifold system.

8. Conclusion

A useful model of the idle state is proposed, which explains the characteristics not only of the ISC system but also of the idle stability. Moreover, it gives some effective method to improve both the ISC system and the idle stability as shown in Table 12. It also provides us with a physical image of the ISC system and the idle stability.

Thus, the model is very useful although it is rather simple and linear at present. Such a model is sure to be useful in realizing ICV.

CHAPTER 3

A Fuel Transportation Model and Its Application to the Analysis of Idle Stability in SPI System

ABSTRACT

In this chapter, both the experimental method and results of the fuel transportaion characteristic are explained first. Then, a simplified evaporation model from the liquid fuel on the wall of the manifold is formulated. Then we explain the aspects of the idle stability in the SPI system using the model. Finally, as another application of the model, a method of the transient fuel control to remove the torque reduction by the lack of the fuel in the cylinders. The author believes the model is useful to realize ICV as well as the proposed model of the idle.

1. Introduction

In the SPI system the fuel is injected from above the throttle valve, and flows partly as an airborne fuel and partly as a liquid fuel on the wall of the manifold. Therefore, the transportation characteristics of the fuel greatly affect the bevavior of the air-fuel ratio, especially in the transient state with a rapid change of the throttle openning. In addition, it is said that these characteristics affect the idle stability in the SPI system. Thus, the SPI system has an advantage that it needs only one fuel injector though it requires more sophisticated fuel control than the MPI system.

The fuel transportation characteristics in the SPI system have been studied by C.F.Aquino[40] and many others.[41]-[43]. Iwano et.al.[44] reported on the behavior of the liquid particle of the fuel in bended manifolds. Andoh et.al.[45] investigated the characteristics by using the visualization technique. C.F.Aquino[40] showed that the characteristics were represented by the first-order lag, and using his model he formulated the necessary additional fuel quantity to keep the air-fuel ratio at a desired value independent of the change in the throttle openning. Also he showed the validity of his formulation by applying it to an actual engine. Andoh et.al.[45] and Inoue et.al.[46] proposed a method of transient fuel control.

Also, several means for transient fuel control were proposed. However, no studies have been reported in detail on the heat and mass transfer problems. Thus, it is necessary to study theoretically the behavior of important parameters in the transport characteristics (e.g., the behavior of the time constant of evaporation of liquid fuel on the wall of the intake manifold) since, with the engine having a discrete effect due to the stroke, the measurement of the transport characteristics of the manifold itself through the engine is essentially impossible.

In this chapter, we show that the transport characteristics are represented by a first-order lag and lead which can be measured with a burner.[42] Then, from the experimental results a quantitative model of fuel transport is proposed[42] and its applications to idle stability are shown.[43]

2. Measurements of the transport characteristics

2.1. Measurement system

The experimental apparatus is shown in Figure 40. The fuel was injected through an injector at a repetition rate with a constant pulse width. The air-fuel ratio was measured by a flame rod above the burner cooled by air. The burner was equipped with a model manifold whose construction is shown in Figure 41. The maximum fuel flow of the burner was 15,000 Kcal/hr, which corresponds to a twentieth of the maximum fuel flow of a 2 litre engine. From the point of load, the burner load corresponds to the wide open throttle. Since the measured values distribute around the average value, they were averaged for 600 samples.

By this apparatus, both the step response and the frequency characteristics of the air-fuel ratio against the fuel flow were measured.

The fuel transport characteristics of the manifold were measured

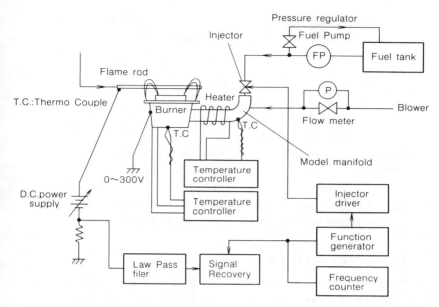

Figure 40. Experimental apparatus of transportation characteristics

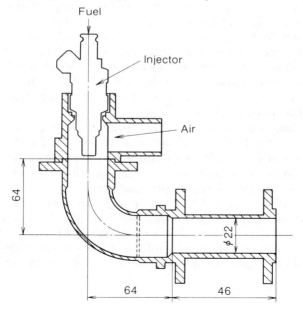

Figure 41. Configuration of model manifold

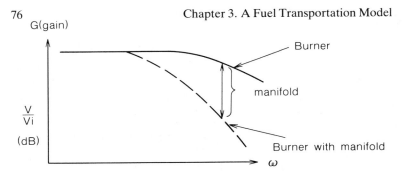

Figure 42. Measurement principle of transportation characteristics of manifold

from the difference in the frequency characteristics of burners equipped with and without the manifold (Figure 42).

2.2. Results and discussions

Figure 43 shows the result on the burner. In the experiment, propane was used as a fuel, because it is difficult to measure accurately the vaporization process of gasoline. In Figure 43(a), the upper trace shows the step response for the burner equipped with the model manifold and the lower that for the burner without the model manifold. Comparing these figures, no differences is found except for the dead time τ_d. Since the dead time agrees with the calculated value from the air velocity in the model manifold, the frequency characteristics about the gain about the burner with and without the manifold must agree. These frequency caracteristics are shown in Figure 43(b).

The white circles in Figure 43(b) represent the frequency characteristics for the burner, and the black circles that for the burner with the model manifold. Apparently, no difference can be found. This result agrees with the above prediction, since, for the gaseous fuel, the manifold acts as a simple dead time.

The step responses of the burner with the model manifold for the gasoline fuel are shown in Figure 44. The temperature T_w represents the temperature of the model manifold. The rise time of the step responses decreases with increasing temperature. This result can be explained qualitatively as follows. If the time con-

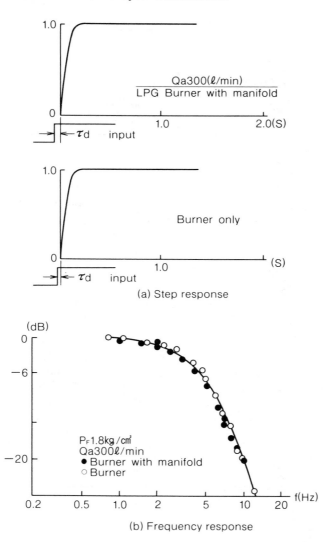

(a) Step response

(b) Frequency response

Figure 43. Step responses and frequency characteristics of burner itself (propane)

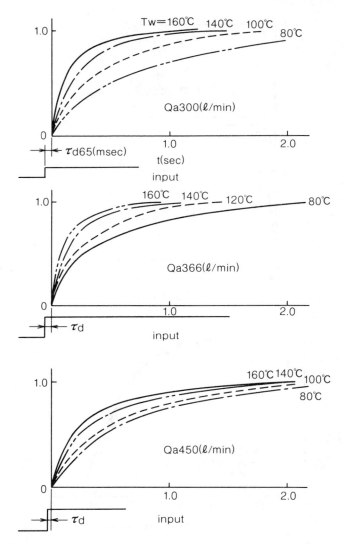

Figure 44. Frequency characteristics of burner with model manifold (gasoline)

stants are determined by the evaporation rate from liquid fuel on the wall of the manifold, the evaporation rate increases naturally with increasing wall temperature. Thus, the time constant becomes short with increasing wall temperature.

Figure 45 shows the results obtained by taking the difference in the frequency characteristics of the burners with and without the model manifold under the principle shown in Figure 42. The following points should be noted: (1)The gain in lower frequencies is almost constant at 0dB. (2)In intermediate frequencies, the gain decreases with increasing frequency. (3)The gain has a certain constant value in higher frequencies. These frequency characteristics are expressed by the following first order lag and lead of $(1+SK_v\tau_m)/(1+ S\tau_m)$ shown schematically in Fig.46 except for the dead time. The time constant τ_m represents the contribution of the evaporation from liquid fuel. The physical meaning of the coefficient K_v is the ratio of the fuel flow carried directly into the burner by the air flow to the injected fuel flow. As is well known, the first-order lag and lead has useful properties. That is, in Figure 46, the lower break point frequency gives the time constant τ_m and the upper the value of K_v. Therefore, approximating the characteristics of Figure 45 to that of Figure 46 we can easily get these values. Plotting these values of τ_m and K_v against the wall temperature T_w, we obtain Figure 47 for the time constant and Figure 48 for the air transportation rate K_v. From Figure 47, it is found that the time constants are strongly dependent on the wall temperature (T_w) and weakly dependent on the velocity of air through the model manifold. In the above experiments, the air velocity was changed by varying the air-flow rate Q_a into the burner. At 450 l/min, the temperature dependency of the time constant becomes weaker. This might be attributed to the change in the experimental condition; the faster air velocity might have changed the flow pattern due to the bending portion in Figure 41 and produced a remarkable change in the formation of liquid fuel.

According to Figure 48, the parameter K_v shows nonzero value and some temperature dependency with weaker dependency on the air velocity. This temperature dependency suggests that the fuel with boiling points below the wall temperature evaporates instantaneously upon reaching the surface of the model manifold and flows with air. On the other hand, the weaker dependency on

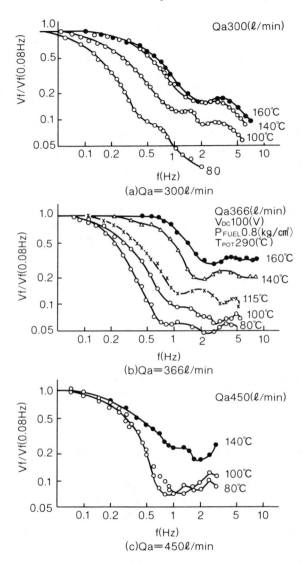

Figure 45. Frequency characteristics of model itself

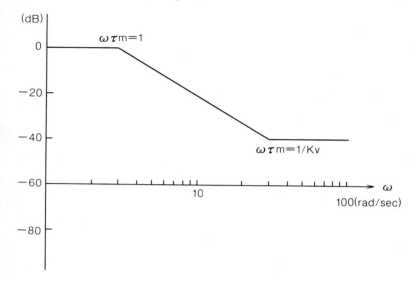

Figure 46. Schematical frequency characteristics of first-order lag and lead

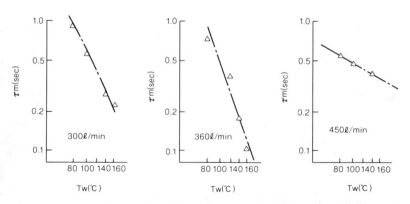

Figure 47. Behavior of time constant τ_m against wall temperature

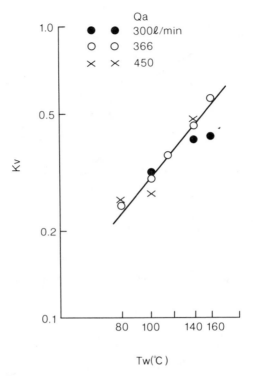

Figure 48. Behavior of K_v against wall temperature

the air velocity suggests that the range is too narrow to show a strong dependency.

After all, the above results strongly suggest that the fuel film on the surface evaporates and the evaporated fuel flows with air. Thus, a simplified evaporation model is proposed in the next section

3. Formulation of the evaporation process

Figure 49 shows schematically the situation of our experiments; the hatched portion shows the liquid film formed on the internal surface of the model manifold.

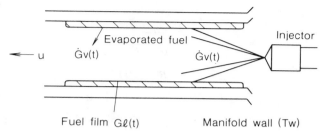

Figure 49. Fuel film model with evaporation from film

Now, the following equations are obtained:

$$\frac{dG_v}{dt} = M_g k_m A_s \frac{P_s(T_l)}{P_b(t) - P_s(T_l)} \qquad (3\text{-}1)$$

$$k_m = \frac{D_v P_b(t) S_h}{R T_a D_l} \qquad (3\text{-}2)$$

$$\dot{G}_l = \dot{G}_i - \dot{G}_v \qquad (3\text{-}3)$$

$$C_{pl} \frac{d}{dt}(G_v T_l) = \frac{\lambda_l A_c}{d}(T_w - T_l) - H_l \dot{G}_v$$

$$+ C_{p1} \dot{G}_i(T_f - T_l) \qquad (3\text{-}4)$$

The first equation expresses the mass transfer from the fuel film and the second the mass transfer rate. These two equations were derived by Hiroyasu[47] for the evaporation from small liquid particles. The authors suggested that the same equations are valid also for the evaporation from the liquid film except for the experimentally determined value of the factor D_l. The third equation represents the mass conservation law of the fuel and the final equation the energy conservation law about the fuel film. The meanings of terms in the fourth equation are as follows. The term in the left hand side represents the time variation of the enthalpy in the film; the first term is the enthalpy from the wall into the film, the second term the dissipated enthalpy by the evaporation, and the last term the enthalpy obtained from the injected fuel. The pertinent equations are given below.

$$A_s = KA_c \tag{3-5}$$

$$d = h/2 \tag{3-6}$$

$$A_c = G_l/(\rho h) \tag{3-7}$$

$$S_h = 0.024\, R_e^{0.8}\, (v/D_v)^{0.43} \tag{3-8}$$

$$\log P_s(T_l) = (9\, T_l/5 + 216.6) \tag{3-9}$$

$$D_v = 0.0505\, (T_d/273)^2\, (760/P_b(t)) \tag{3-10}$$

For simplification, the following assumptions were made: (1)The diffusion is caused by the concentration gradient and the temperature distribution in the fuel film is negligible. (2) The wall temperature and the thickness of the liquid film are not varied during the evaporation. (3) The heat transfer from the film to the air flow and the change in the density of the film are negligible.

In these equations, The notation A_s represents the effective area of the liquid film, h the thickness of the film, the fourth equation the Sherwood number, the fifth the saturation vapor pressure. Assuming a value for h, all necessary values can be determined from Eq.(3-1) to Eq.(3-10). Referring to the results of Sawa[48], we assumed $h=0.2$mm, and we carried out the simulation of the step response for the stepwise change in the fuel flow, using the above equations. As a result, Figure 50 was obtained, which shows the responses of the fuel evaporation per unit time and the weight of the fuel film when the fuel flow rate is altered stepwise from 0.3 to 0.5 g/s. From this response of the evaporation rate of fuel flow, the time constant can be easily determined theoretically. In Figure 51 the results obtained by simulation are shown together with the values estimated from the results of Aquino.[40] Except at 450 l/min, the results show a similar trend in the results of Aquino and the present study. This shows the validity of the above formulation. The parameter K_v is not included in the model, because K_v is determined by various factors and is difficult to include in the above simple model.

A comment on the thickness h of the fuel film assumed in the simulation will be given below. By simulation, it was found that the thickness affects greatly the time constant of the evaporation; the time constant increased with increasing thickness. When the

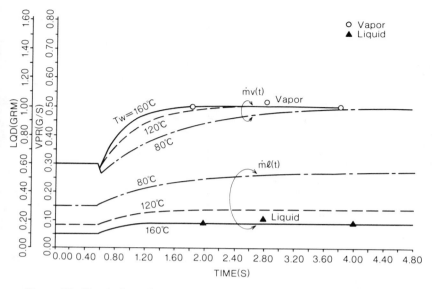

Figure 50. Simulation of A/F behavior against stepwise change of fuel-flow rate

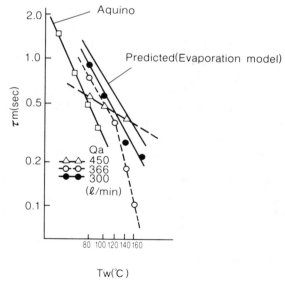

Figure 51. Comparison of time constant τ_m

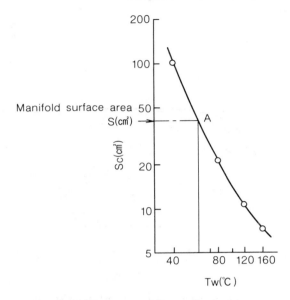

Figure 52. Film area against wall temperature

wall temperature decreases, the weight of the fuel film increases and, accordingly, the surface area of the film increases. Thus, when the wall temperature is decreased gradually, the surface of the manifold wall will be covered totally with the fuel film at a certain temperature T_{wo}. In such a situation, some gasoline must leak from the junction between the burner and the model manifold. Figure 52 shows the situation: the area of the fuel film predicted by assuming $h=0.2$mm is plotted against the wall temperature T_w.

We estimated the critical temperature T_{wo} from our experimental condition to be 60 °C as shown by the point A in Figure 52. The observed temperature was 70 °C. Thus, the assumption of $h=0.2$mm is rather correct. Further, it should be noted that the analytical formulation is rather effective in understanding the phenomenon of fuel transportation.

4. Simulation of the hunting phenomenon of the idle speed in the SPI system

It seems apparently possible to simulate the behavior of the idle speed by combining the transport model described in the preceding section with the engine model as shown in Figure 53, which is composed of the fuel transport block, and the load-inertia block. The calculations of the engine torque obtained by the combustion and the engine friction were carried out by the following equations proposed by Awano.[49]

$$T_i = \frac{10}{4\pi} P_i (1000 V_h) \tag{3-11}$$

$$T_{fy} = \frac{10}{4\pi} P_f (1000 V_h) \tag{3-12}$$

$$P_i = \frac{\phi(\lambda)K(\epsilon)}{420 + t_c} P_b (1 - 0.82 \frac{P_r}{P_b}) + 0.0136 (P_b - P_r)$$

$$P_f = (0.00296 - 5.65 \times 10^{-7}N) P_b - 0.79 + 9 \times 10^{-4}N + P_{fo}$$

$$T_b = T_i - T_{fy} \tag{3-13}$$

Euler's equation was used for the determination of the engine speed.

$$\frac{\pi}{30} J \frac{dN}{dt} = T_b \tag{3-14}$$

The calculation procedures used were as follows:

(1) By assuming the thickness of the fuel film h to be 0.2 mm, the manifold absolute pressure P_{bo} to be 250 mm Hg, and the idle speed N_o to be 650 rpm, Eqs. (3-1), (3-2), (3-3) and (3-4) were solved simultaneously. Thus, we obtained the fuel-flow rate by the evaporation from fuel film and the film temperature T_{lo} at equilibrium.

(2) Setting an appropriate value for the base engine friction P_{fo}, we calculated the instantaneous engine speed from Eqs. (3-11) to (3-14).

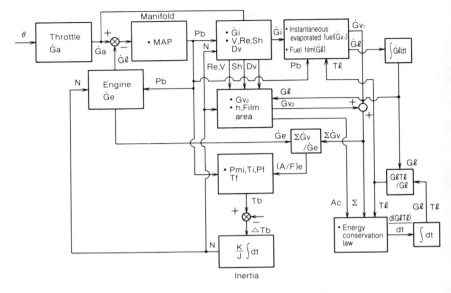

Figure 53. A block diagram for simulating engine system

As a result, we were able to obtain two kinds of infomation, one of which is on the response of the engine speed against any disturbance. For example, we calculated the step response against the stepwise torque variation. The other is on the frequency characteristics of various quantities in the system; by opening the loop at the output of the load-inertia system and separating the input N into the manifold system from the output N, we calculated the variations in physical quantities such as the air-fuel ratio, manifold pressure, the brake torque, and the output engine speed against the sinusoidal variation in the engine speed. Further, we obtained the frequency characteristics by calculating these variations for various frequencies of the input N.

Figure 54 shows an example of the response of the engine speed against the stepwise disturbance of the load torque. The A/F ratio represents the aimed air-fuel ratio. According to Figure 54, the idle stability is poor at larger A/F ratios; at $A/F=18$ the engine

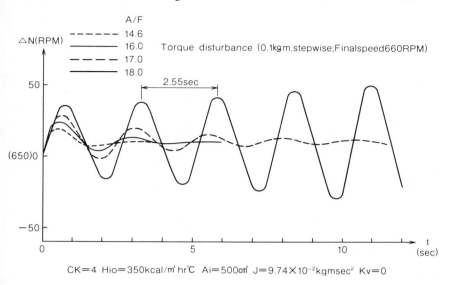

Figure 54. A simulation result of step response under torque disturbance

speed oscillates and the amplitude grows gradually with time. The frequency of the oscillation was found to be about 0.4 Hz, which is very close to that observed often with actual engines; this calculation method simulates well the hunting of the engine speed. However, we cannot explain the cause of the hunting by Figure 54. Thus, it is necessary to calculate the frequency characteristics by the method mentioned above. The result obtained for the situation under oscillation is shown in Figure 55. The phase difference between the total torque variation T_b and the original variation of the engine speed is $\pi/2$; thus, the phase condition discussed in Chapter 2 is satisfied and the system causes a spontaneous oscillation. Also, the variation in $\Delta(A/F)$ occurs in advance of the variation in the manifold absolute pressure ΔP_b; assuming the manifold volume to be zero, the phase difference between ΔN and ΔP_b is 180 degrees. Therefore, the phase of the variation in the air fuel ratio $\Delta(A/F)$ is in the third quadrant. Schematically drawing this phase relation, we obtain a vector diagram shown in Figure 56, which is quite

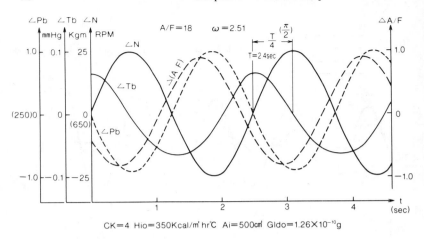

CK=4 Hio=350Kcal/m²hr°C Ai=500cm² Gido=1.26×10⁻¹⁰g

Figure 55. Phase relations between $\Delta N, \Delta P_b, \Delta P_b$, and $\Delta(A/F)$

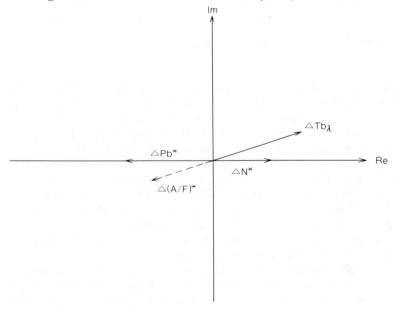

Figure 56. Schematical vector diagram of $\Delta N, \Delta P_b$, and $\Delta(A/F)$

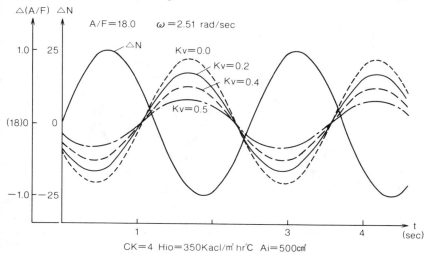

Figure 57. Behavior of A/F variation against K_v

similar to that discussed in the preceding chapter and helps us understand intuitively the aspect of the generation of the hunting phenomena in SPI. It is obvious that the variation vector plays an important role in the generation of idle speed hunting.

We examine next the behavior of A/F variation against the parameter K_v and the aimed air-fuel ratio. The K_v dependency on the A/F variation is shown in Figure 57; when K_v increases, the amplitude of the variation decreases though the phase relation remains unchanged. This result agrees well with the real aspects of idle speed hunting.

The behavior of the A/F variation for the aimed air fuel ratio is shown in Figure 58. The amplitude of the A/F variation increases at leaner aimed air-fuel ratios, which agrees well with the fact that the oscillation amplitude of the idle speed increases at leaner aimed A/F ratios.

Thus, the simulation method proposed in this study is very useful to the analysis of the idle stability in SPI system. However, this method is too complex for us to understand intuitively the phe-

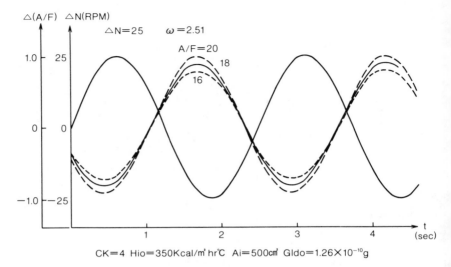

CK=4 Hio=350Kcal/m² hr°C Ai=500cm² GIdo=1.26×10⁻¹⁰g

Figure 58. Behavior of A/F variation against aimed A/F

nomenon of idle stability. Therefore, in the next section, a simplified treatment based on the present model is proposed.

5. Simplification of the evaporation model

The analytical model, though effective, is complex; for example,
taking a look at the Eqs. (3-1), (3-2), (3-3), and (3-4), it cannot be
understood that the frequency characteristics of the fuel transportation is represented by the first-order lag. In this section, more
simplified equations will be derived. Considering the small variations of all the variables in Eqs. (3-1)–(3-4) from the equilibrium,
we finally obtain the following equations.

$$\Delta \dot{G}_v^* = \frac{1}{F(S)} \left[1 + k\frac{\tau_2}{\tau_1} + S \left(1 - 2k + k\frac{T_f}{T_{lo}} \right) \right] \Delta \dot{G}_i^*$$

$$- \frac{P_{so}}{P_{bo} - P_{so}} \frac{S\tau_1(1 + S\tau_2)}{F(S)} \Delta P_b^* \qquad (3-15)$$

$$F(S) = S^2\tau_1\tau_2 + S\{\tau_2(1 - k + H_l/C_{pl}T_{lo}) + \tau_1\} + 1 + \tau_2/\tau_1 \quad (3\text{-}16)$$

$$\tau_1^{-1} = \frac{\lambda_t}{C_{pl}d \cdot \rho h}(T_w/T_{lo} - 1) \quad (3\text{-}17)$$

$$\tau_2^{-1} = \frac{\lambda_t}{C_{pl}d \cdot \rho h} + \tau_1^{-1} \quad (3\text{-}18)$$

$$\tau_l = G_{lo}/\dot{G}_{io} \qquad k = \alpha\{1 - P_{so}/(P_{bo} - P_{so})\} \quad (3\text{-}19)$$

$$\alpha = H_l/ART_{lo} \quad (3\text{-}20)$$

where we assumed that the fuel was composed of a single component such as octane (Eq.(3-20)).

The first equation is most important; the term in the left-hand side represents the variation of the evaporation rate of fuel flow. The first term in the right-hand side expresses the contribution of the variation in the injected fuel flow to the evaporation fuel flow and the second term represents that of the manifold absolute pressure ΔP_b to the evaporation fuel flow. In the above, we assumed for simplification $K_v = 0$ and the mass transfer coefficient k_m has a constant value determined by the equilibrium values of all variables. Generally, a small variation of air fuel ratio is given by the following equation.

$$\frac{\Delta\lambda}{\lambda_o} = \frac{\Delta\dot{G}_a}{\dot{G}_{ao}} - \frac{\Delta\dot{G}_v}{\dot{G}_{vo}} \quad (3\text{-}21)$$

According to Eqs. (3-15) and (3-21), we find two interesting points.

(1) Under the experimental conditions of this study, the step response of the air fuel ratio for the stepwise change in the fuel flow is given by a coefficient of $\Delta\dot{G}_i/\dot{G}_{io}$, because $\Delta\dot{G}_a = 0$ and $\Delta P_b = 0$:

$$\frac{\Delta\lambda}{\lambda_o} = \frac{-1}{F(S)}\left[1 + k\frac{\tau_2}{\tau_1} + S\left(1 - 2k + k\frac{T_f}{T_{lo}}\right)\right]\frac{\Delta\dot{G}_i}{\dot{G}_{io}} \quad (3\text{-}22)$$

It is clear that the coefficient reduces essentially to the first-order lag, because the numerator has a first order polynomial about S while the denominator $F(S)$ has a second-order polynomial. In the actual engine with both a small manifold volume and a

SPI system, we often meet the situations that only the fuel flow rate is changed while the throttle opening and the manifold pressure are kept constant. In these cases, it is generally considered that the characteristics of fuel transportation can be approximated by the first-order lag

(2) If only the manifold absolute pressure is varied and both the injected fuel and the air-flow rate are not varied, $\Delta \dot{G}_i = 0$ and $\Delta \dot{G}_a = 0$. Then, the variation of the vaporized fuel-flow rate is given by the second term in Eq. (3-15). In this case, the variation of the air-fuel ratio is given by the equation:

$$
\begin{aligned}
\frac{\Delta \lambda}{\lambda_o} &= \frac{P_{so}}{P_{bo} - P_{so}} \frac{S\tau_1(1 + S\tau_2)}{F(S)} \frac{\Delta P_b}{P_{bo}} \\
&= -\frac{P_{so}}{P_{bo} - P_{so}} \frac{S\tau_1(1 + S\tau_2)}{F(S)} \frac{\Delta N}{N_o}
\end{aligned} \tag{3-23}
$$

In this situation, the characteristics of the fuel transport are effectively represented not by the first-order lag but by the zero-order as shown by Eq. (3-23). The second term in Eq.(3-15) represents the effect of the variation of the MAP on the evaporation rate, and plays an important role in the idle stability in the SPI system as will be described later in detail. Since in the idle state both the air-flow and and the fuel-flow rate are not changed, the second term determines the behavior of the air-fuel ratio.

It should be noted that the characteristics of the fuel transport behave differently in accordance with the situation. This is one of the most important conclusions obtained from the simplified model.

6. Application of the model to the explanation of the idle stability in the SPI system

In this section, we explain the phenomenon of idle stability in the SPI system using the vector loci proposed in the previous section. According to the previous chapter, the idle stability in the SPI system must be good because the manifold volume in the SPI system is generally smaller than that in the MPI system. This reduces the phase lag of the torque variation vector due to the variation of the

MAP, and it must give a good idle stability. However, this is not always valid in the SPI system, because the variation of the air-fuel ratio due to the variation of the evaporated fuel is generated and it increases the torque variation. Therefore, even with the MPI system, we must take into account the variation of the air-fuel ratio when liquid fuel film exists in the manifold wall as shown later.

Now, we try to explain the phenomenon of idle stability in the SPI system by using the above results.

6.1. Vector loci of the torque variation induced by variation of the air-fuel ratio and the vector diagram in the SPI system

We examine first the phase characteristics of the variation in vector given by Eq. (3-23). Figure 59 shows the phase characteristics with various frequencies. Without a negative sign, at $\omega = 0$, the phase of the demoniminator is equal to zero and that of the numerator to $\pi/2$. On the other hand, at $\omega = \infty$ they have the same values π. Therefore, without a negative sign the phase of the equation ranges from 0 to $\pi/2$. This means that the variation vector of the air-fuel ratio exists in the third quadrant as shown in Figure 59 quite similar to Figure 56. Since $K_\lambda < 0$ in the region where the air-fuel ratio of the equilibrium is leaner than 13, the vector of the torque variation induced by that of the air-fuel appears in the first quadrant. Combining this result and the previous vector loci, we obtain Figure 60. At a glance of Figure 60(a), one could understand that the torque variation $\Delta T_{b\lambda}$ induced by the air-fuel ratio makes the total torque variation ΔT_b larger and that it makes the idle stability in the SPI system worse. In other words, the two torque variation ΔT_{bp} and $\Delta T_{b\lambda}$ determine the aspects of the idle stability in the SPI system. Thus, in the SPI system, the idle stability is often poor even though the manifold volue is very small. This result is in a remarkable contrast to the result of the MPI system. This situation is shown in Figure 60(a) and (b). According to Figure 60(a) the torque variation $\Delta T_{b\lambda}$ induced by the the variation of the air-fuel ratio makes the idle stability worse in the SPI system. On the other hand, the torque variation $\Delta T_{b\lambda}$ induced by the derivative fuel modulation, for instance, makes the idle stability better in MPI system.

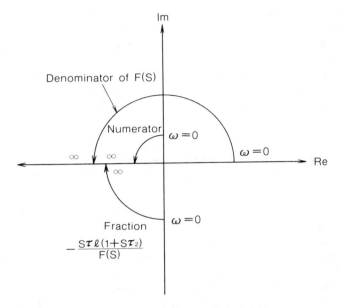

Figure 59. Phase characteristics of $\Delta\lambda$

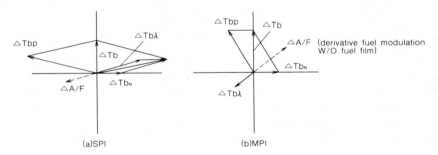

Figure 60. Schematical vector diagram of both SPI and MPI

6.2. Influence of the fuel film on the idle stability in the MPI system

In the above section the influence of the fuel film on the idle stability in the SPI system was shown. Against this, in this section,

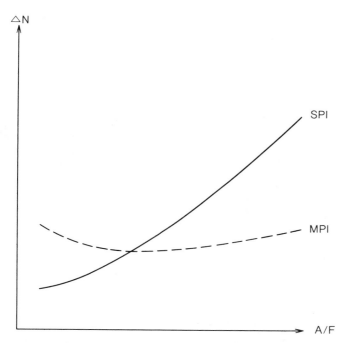

Figure 61. Schematical behavior of idle stability against A/F

the influences of the fuel film in MPI system are explained in detail. For this purpose we review first the differences of the idle stability between the SPI and MPI system.

As shown in Figure 61, the idle stability gets worse when the air-fuel ratio at equilibrium becomes leaner. This fact was explained by simulation earlier in this chapter. In this section, we try to explain this intuitively by a vector diagram. The coefficient K_λ that multiples the variation of the air-fuel ratio generally increases with increasing equilibrium air-fuel ratio as shown in Figure 62, because K_λ is essentially given by the derivative of the mean effective pressure against the aimed air-fuel ratio. This means that the vector of the torque variation ΔT_b increases with increasing aimed air-fuel ratio. According to Figure 60(a), it is obvious that the increase of

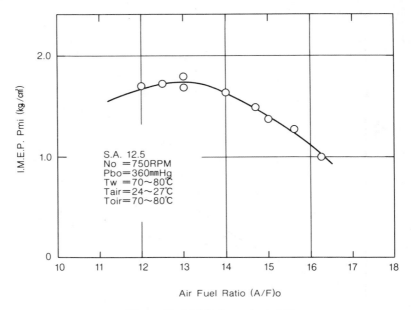

Figure 62. I.M.E.P. against A/F

K_λ results in a larger total torque variation ΔT_b, and, accordingly, increases the loop gain and makes the idle stability poor. On the other hand, if the aimed air fuel ratio becomes richer than about 13, the coefficient K_λ becomes positive as shown in Figure 62. This inverts the direction of the vector $\Delta T_{b\lambda}$ and, as a result, the total torque variation becomes smaller. Thus, the loop gain becomes smaller and makes the idle stability better. In this manner, the behavior of the idle stability in SPI system against the aimed air-fuel ratio can be explained.

On the other hand, in the MPI system (D system), there is little influence of the torque variation of the air fuel ratio as previously mentioned; the behavior against the aimed air-fuel ratio becomes independent. This is also shown in Figure 61.

As mentioned above, in the SPI system the existence of the fuel film greatly affects the idle stability. If this result is truly correct,

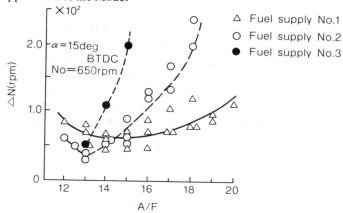

Figure 63. Relation between fuel-film formation and idle stability

the existence of the fuel film must influence the idle stability also in the MPI system. In this meaning Nogi et al.[50] report experimental results of great interest.

They carried out experiments on the idle stability by varying both the location and the atomization of the fuel injector with a single fuel control logic (L-system). Their results are summarized in Figure 63. The black circles (No. 3) show the behavior of the idle stability against the aimed air-fuel ratio when the fuel in the film state was injected at 15 cm upstream of the intake valve of the engine. The white circles (No.2) show the same behavior observed when the fuel was supplied in a well-atomized condition at the same location as with the black circles. The last case (No. 1) of the triangle shows the behavior of the usual case; the fuel was injected at the nearest position toward the intake valve. Therefore, it is inferred that the quantity of the fuel film decreases in the order: No.1, No.2 and No.3. Comparing Figure 63 with Figure 61, one can see that, with increasing fuel film, the behavior of the idle stability in the MPI system comes closer to that in the SPI system.

Strictly speaking, the experimental results by Nogi et al. are for the MPI system with an air-flow sensor (L system). As shown in Eq. (2-48), the spontaneous fuel modulation similar to the derivative fuel modulation was carried out in L system. Therefore, we

explain the influence of the fuel film in the MPI system on the idle
stability, taking into acocunt the modulation effect represented by
Eq. (2-13). Here, we explain the influence of the fuel film by
calculating the loop again, because the effectiveness of the vector
diagram has been shown completely and because the vector diagram for the idle stability of the present case is rather complex.
The following equations are derived in the previous manner.

$$\Delta T_{bp} = - \frac{K_p}{1 + S\tau_a} \Delta N^* \tag{3-24}$$

$$\Delta T_{b\lambda} = K_\lambda \left\{ - S\tau_a \Delta P_b^* - \frac{1 + SK_v\tau_m}{1 + S\tau_m} (\Delta P_w^* + \Delta N^*) \right\}$$
$$- K_\lambda \frac{P_{so}}{P_{bo} - P_{so}} \frac{S(1 - K_v)\tau_m}{1 + S\tau_m} \Delta N^* \tag{3-25}$$

where, we made the following approximation:

$$\frac{S\tau_1(1 + S\tau_2)}{F(S)} = \frac{S(1 - K_v)\tau_m}{1 + S\tau_m} \tag{3-26}$$

The third term on the right-hand side of Eq. (3-25) represents the
contribution of the fuel film to the variation of the air-fuel ratio.
As in the same manner, we can calculate the loop gain. Figure 64 is
an example of the result. Apparently, with smaller values of K_v,
the behavior against A/F of MPI comes closer to that of SPI system
and to Figure 63. That is, the result explain the experimental result. The fuel film also makes the idle stability worse in the MPI
system. Also shown in Figure 64 a different behavior is observed
for K_v=0.9. This would suggest that even in MPI system with fuel
injection toward the intake valve there is a little fuel film.

From the above statement, it is obvious that the better the atomization of the fuel injector the better the idle stability. Moreover, we
can get the better combustion by improving the atomization. It has
been said that the particle size is controlled by the atomization in the
narrow gap between the throttle valve and the throttle bore in the
SPI system. However, Kubo[54] showed that it was controlled by the
atomization of the fuel injector by his experiments on the particle
size at the locations above and below the throttle valve. Therefore,
it is important to install injectors with better atomization.

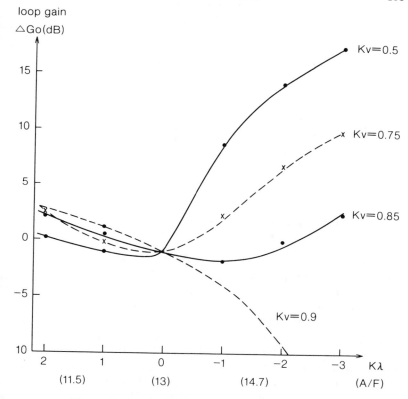

Figure 64. Relation between loop gain (ΔG_o) and K_v

7. Application of the model to the derivation of the additional fuel quantity in the transient state

The present model is also applicable to the derivation of the additional fuel quantity in the transient state for small variations of various variables from equilibrium. An example of application to rapid acceleration will be given here.

Consider the case where the transfer function is represented by the first order lag and lead $(1 + SK_v\tau_m)/(1 + S\tau_m)$ as shown in Figure 65. In this figure, $\Delta I_b(S)$ represents the base fuel-flow rate,

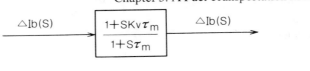

(a) Steady state

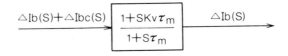

(b) Transient state

Figure 65. Block diagram for transient fuel control

Apparently, in the steady state, the output is also given by $\Delta I_b(S)$. However, the output differs from the input $\Delta I_b(S)$ in the transient state because of the delay by the fuel transportation. The role of the additional fuel in rapid acceleration is to complement the fuel flow by the additional fuel flow and keep the output level $\Delta I_b(S)$ during the transient period. Therefore, the formulation of the additional fuel flow is to derive the additional flow $\Delta I_{bc}(S)$ which makes the output equal to $\Delta I_b(S)$ during acceleration. Therefore, the derivation of the additional fuel flow is carried out in the following procedures. The output becomes equal to $\Delta I_b(S)$ when the input $(\Delta I_b(S) + \Delta I_{bc}(S))$ is applied to the system. Thus, the following equation is derived.

$$\frac{1 + SK_v\tau_m}{1 + S\tau_m}\{\Delta I_b(S) + \Delta I_{bc}(S)\} = \Delta I_b(S) \tag{3-27}$$

Solving this equation about $\Delta I_{bc}(S)$, we obtain Eq. (3-28).

$$\Delta I_{bc}(S) = \frac{S(1 - K_v)\tau_m}{1 + SK_v\tau_m} \Delta I_b(S) \tag{3-28}$$

This equation represents the additional fuel-flow rate $\Delta I_{bc}(S)$. Since we cannot create an image of the equation only by seeing Eq. (3-28), we draw a pattern given by the equation for the stepwise variation of the base fuel flow rate $\Delta I_b(S)$. Figure 66 shows the result. It

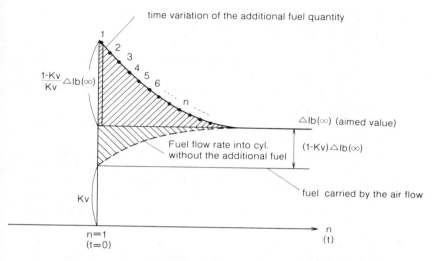

Figure 66. Time variation of additional fuel flow

should be pointed out that the pattern of the additional fuel-flow rate for the time agrees well with our experiences. In the actual case, the determination of the additional fuel-flow rate is not so easy, because there are many additional factors to be considered, such as the contribution of the sampling effects of the engine stroke, the delay in the air detection, the processing time in the processor, and the arrival time of the fuel carried by air. These additional factors contributing to the transient fuel control are explained in the literature[52]–[54] listed at the end of this book. However, the pattern shown in Figure 66 represents the most basic aspect of the additional fuel-flow rate. Thus, our model is also useful to transient fuel control.

8. Conclusions

We carried out experiments with a burner and the formulation of the fuel transportation to explain the characteristics of the idle stability in SPI and MPI system. The results obtained are:

(1) The characteristics of fuel transport in the manifold is given by the first-order lag and lead $(1 + SK_v\tau_m)/(1 + S\tau_m)$.

(2) The parameter K_v represents the ratio of the fuel carried by air to the total fuel, and the time constant τ_m the delay caused by evaporation from the fuel film.

(3) The formulation of evaporation from fuel film explains well the temperature dependency of the time constant.

(4) The simulation based on the present model explains well several aspects of idle stability in SPI system.

(5) Simplification of the formulation resulted in a simple equation for the variation of the air-fuel ratio caused by the fuel film.

(6) The simplified equation revealed that the variation of the air ratio under the existence of the fuel film was controlled by the variation both of the injected fuel and of the manifold pressure.

(7) From the simplified equation, the vector diagram was obtained which explained well both the dependency of the air-fuel ratio on the idle stability and the cause of the large variation of the idle speed.

(8) Fuel-film thickness probably affects the performance characteristics of the MPI system to some extent.

CHAPTER 4

Status of Data-Driven Processor in Japan

ABSTRACT

New type processors are essential for the development of the intelligently controlled vehicle (ICV) in which an immense amount of processing is required. We have to watch the trends in microprocessors. In this chapter, the present status of new type processors is briefly reviewed.

1. Introduction

Microrocessors are now widely used in various fields calling for high-level processing quality and quantity. A great increase in the calculation rate is desired by users. Software engineers also look for high-speed microprocessors to use high-level languages instead of machine language.

Generally, there are three approaches to increasing the calculation rate of microprocessors: (1) use high-speed materials such as GaAs; (2) use the most suitable architecture for high-speed calculation; and (3) increase the bit length to process the data. The third approach is the current trend; the bit length has been increasing from 4 to 32 bits along with progress in microelectronics technology. On the other hand, the disadvantages of the von Neumann architecture have been pointed out and several new architectures proposed.

In this chapter, we introduce briefly the present situation of research on a processor with a data-driven architecture and the language to be used.

2. von Neumann and non-von Neumann architectures

A large number of processors adopt the von Neumann type architecture, in which the calculation process is carried out sequentially.

The actual processor consists of a central processing unit (CPU) and a memory (ROM,RAM) where the data and the program are stored. These are connected with each other by bus lines as shown schematically in Figure 18. (See Figure 18) In the von Neumann architecture, the calculation is carried out in the following way; for simplicity, we consider the calculation of the following equation.

$$G = A \times B + C \times D + E \times F$$

(1) The program code of LDA is sent to CPU from the memory through the bus after the program counter (PC).
(2) The CPU fetches and decodes the program code of LDA.
(3) Since the code is to load the data A indicated by the address into the A register, the CPU demands to send the data A to CPU.
(4) Actually the data A is sent to CPU through the bus line and stored in the register.
(5) The data B is stored in another B register in the same way.
(6) Then, the machine code of MLY(X) is sent to CPU through the bus line.
(7) CPU fetches and decodes again the code, and performs the arithmetic calculation of the product A × B at ALU in CPU.
(8) This processe is repeated to complete the calculation.

Obviously, the actual operation time of ALU is quite short though the bus line is very busy. Thus, the processing rate is controlled by the time of the transfer process between the memory and CPU; this is called the von Neumann bottle neck. Therefore, in the von Neumann architecture, it is important to increase the transfer rate and the bit length to increase the calculation rate.

Several non-von Neumann architectures have been proposed to overcome the disadvantages of the von Neumann architecture. The data-driven architecture is one of them. Though it was originally proposed in the U.S.A., it is being developed more actively in Japan.

In the data-driven architecture, processing is carried out whenever two data necessary for the calculation (A and B in the above example), meet each other. The stream of the processing is written by the data-flow graph proposed by Dennis as shown in Figure 67,

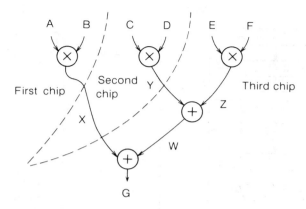

Figure 67. Examples of data-flow graph

while the flow in von Neumann architecture is written by the flow chart shown in Figure 18(a). Comparison of Figure 18(a) and Figure 67 teaches that the data-driven architecture makes much of parallel processing. Thus, the data-driven processor is essentially advantageous for high-speed calculation. The difference between these architectures will be more obvious when a simple processing of the following equation is considered. The flow chart and the data flow graph are shown in Figure 68.

$$U = A \times B + C/D \qquad V = A \times B - C/D$$

Demand-driven architecture is another non-von Neumann architecture. In this architecture the processing is performed in response to the demand. The processing speed of the demand-driven architecture is considered to be slower than that of the data-driven architecture. The driving method of processing is classified in Table 13. The control-driven type of the first column is the von Neumann architecture in which the calculation is performed by a program (control). As shown in Table 13, there are two driving types in non-von Neumann architectures, but we introduce the data-driven architecture because of its greater potential.

The history of the data-driven architecture is summarized in

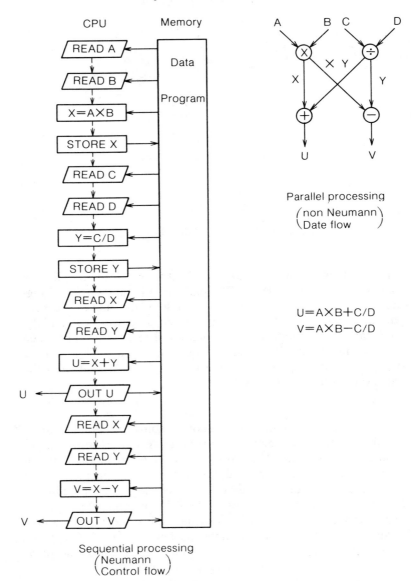

Figure 68. Comparison between flow chart and data-flow graph

Table 13. Classification of driving method.

Driving method	Features
1. Control driven	Processing driven by control program
	Proposed by von Neumann.
2. Data driven	Processing driven by data flow
	Parallel processing
3. Demand driven	Processing driven by processing demand
	Processing rate smaller than that of the data-driven method.

Table 14. In 1976, its R & D was introduced into Japan by Arbind. Some data-driven processors called ImPP were commercialized by NEC corporation in 1983 and 1985.

The most difficult point in the data-driven processor is how to construct the architecture, especially on a single chip. One of the possibilities is the multiprocessor configuration. For example, a configuration composed of four chips would be conceived as shown in Figure 77; the multiprocessor configuration seems suitable for the data-driven architecture, but not for the von Neumann architecture.

3. Realization of the Data-Driven Architecture on a Single Chip[55,56]

As mentioned above, the multiprocessor configuration seems suitable for the data-driven architecture. We consider here how to construct the data-driven architecture on a single chip. Circular pipeline processing was proposed by several investigators, the concept of which schematically is shown in Figure 69. Since the concept of the parallel procssing in the circular pipeline configuration is very similar to that of the mass-production line, we explain first parallel processing in the production line. Figure 70 shows schematically the operation in the production line. The rectangles represent the processes involved. Several materials are thrown in the line as shown by the arrows and the products are derived from the right end. In such a

Table 14. Research history of data-driven processor

Year	Events
1964	CDC6600 a vector machine with a processing pipeline, was commercialized. Single instruction stream
1967	IBM360/91 was commercialized, which used a processing pipeline as well as CDC6600. Basic research on data-driven architecture was started at MIT and Stanford Univ.
1970	Dennis(MIT) studied strenuously data driven architecture, and proposed suitable language for data-driven architecture was proposed.
1976	Research at MIT was introduced to Japan by Arbind.
1978	TI proposed a proto-type processor of DDP (Distributed Data Processor). Four PE (Processing Elements) system.
1983	NEDIPS processor for image processing was commercialized by NEC Corporation.
1985	ImPP (Image Pipelined Processor) was commercialized by NEC.

line, the products are produced at every interval called the tact time. This means that all the processes denoted by rectangles must be carried out simultaneously in a tact time. This is parallel processing. In circular pipeline processing, the same situation holds; in Figure 69, while data are running around the circular pipeline, they undergo processing at the processing steps on the circular pipeline. This suggests the possibility of parallel processing in a single chip. Thus, the construction of the circular pipeline seems very useful to performing parallel processing in a single chip.

The pipeline parallel processing is used recently in the von Neumann architecture and the digital signal processor (DSP). However, if some branches are included in the program, the principle of pipeline parallel processing is not always effective because of misbranches often observed.

The configuration of the execution of the data-driven architecture is shown in Figure 71. In this figure, at the paring process (PP), the data pair necessary for the calculation is found out. The program, which is quite different from that of Neumann architecture as will be shown later, is stored in the program store (PS). The actual calculation is carried out in the arithmetic logic unit (ALU).

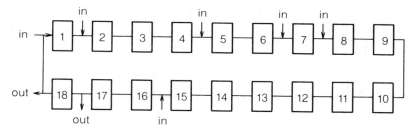

Figure 69. Principle of parallel processing in circular pipeline configuration

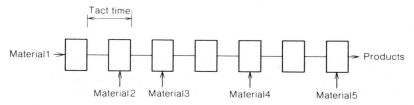

Figure 70. Parallel processing in production line

Now, we explain the actual operation in the circular pipeline processing by means of Figure 72. We consider the calculation of D $=A \times B + C$. The data A, B, and C are thrown into the input-output control portion. Then, the data are reformed into a packet form with a tag and circulate toward the paring process (PP). Examples of the packets are also shown in the right-hand side of Figure 72. The packet is composed of the tag (upper part) and the data (lower part). The tag in the packet is composed of several bits, which represent the node number in the data-flow graph shown in the lower part of Figure 72, operation code, input-output code, and the right-left code. For instance, in packet A, the first number 0 in the tag means the node 0, the operation code * multiplication, the notation IN the input data, and the last notations L and R the code to discriminate the data pair necessary for the processing. Thus, only the last codes are different in the tag of the data packets A and B.

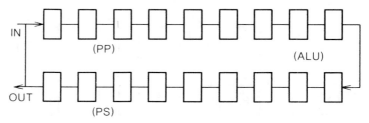

PP : Pairing Process
ALU : Arithmetic Logical Unit
PS : Program Store

Figure 71. Example of circular pipeline

Now, in the paring process, the two packets with different last
code in the tags are found out, and these two data packets are
reformed into one packet as shown in the second packet. In this
process the last code in the tag becomes meaningless, and the data
A and B are combined. When the data packet flows into the ALU,
the multiplication denoted in the second code of the tag is performed and the data A is replaced by the product $W(=A*B)$ and
B is extinguished. This packet flows into the program store (PS).
The stored program in the Program Store, is composed of two
parts. One is the memory of the operation codes and the other that
of the next destination (the node number): the table shown in
Figure 72 is stored in PS and when the packet reaches there the tag
of packet W is rewritten by referring to the table. For example, the
first code is replaced by 1, because the next destination is the node
1 as shown in the data-flow graph. Next, the second code is replaced by the instruction code $+$, because the calculation in the
next node is a summation. The third code is replaced by the code
IN, which means that this packet is not output yet. Lastly, the
fourth code is written as L, because this packet flows into the node
1 from the left-hand side. Thus, the packet flows from the input-
output control portion to PS, and flows back into the input-output
control portion. Since the third code in the tag is IN, this packet
circulates to the paring process (PP) and undergoes the same procedures; in PP the packet W finds out the packet C, the summation is

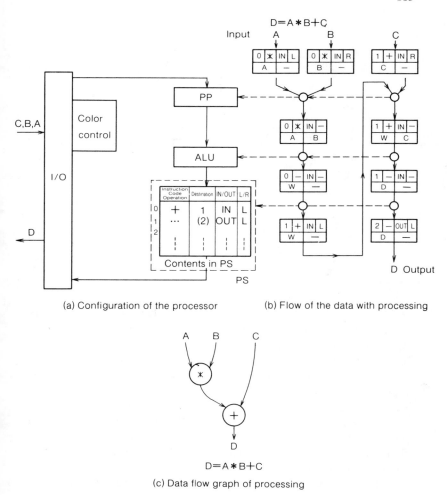

(a) Configuration of the processor

(b) Flow of the data with processing

(c) Data flow graph of processing

Figure 72. Explanation of execution in circular pipeline

performed in ALU, and in PS the third code in the tag is written as OUT. When the packet D circulates again to the input-output control, the packet is decoded and discharged as a processed datum D from the circular pipeline. Other data packets not shown

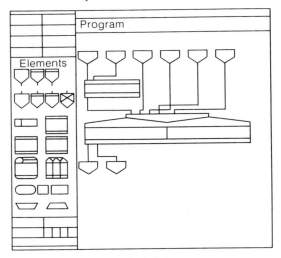

Figure 73. Example of diagrammatic language

in Figure 72 are also flowing in the circular pipeline when the above procedures of calculation are carried out. Therefore, the operation speed and the calculation speed can be much higher. This is also the case with a single chip in which a parallel processing is carried out at an interval required for a single circulation. In this architecture, the calculation is driven by flow of data packets, which resulted in the naming of data driven architecture.

The color control portion controls the data generation, which means formation of a series of data to be calculated. For example, there is a series of sampled data sampled at the same time, which we often encounter in the field of the control.

The data-driven processor is being developed extensively in Japan for its merits described above.

4. Language Used in Data Driven Processor[57][58]

As for the language to be used in the data driven processor, we can think of a language that represents the data-flow graph of desirable

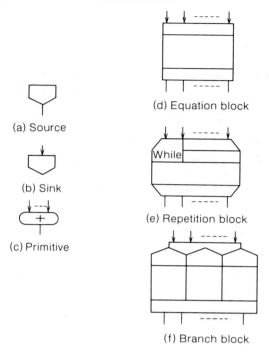

(a) Source

(b) Sink

(c) Primitive

(d) Equation block

(e) Repetition block

(f) Branch block

Figure 74. Example of elements of diagrammatical language

processing, since the data-flow graph is essential both to understand the contents of processing and to carry out the calculation.

As an example, a graphical language is shown in Figure 73,
which is a typical display of a graphic editor. The left-hand side of
this picture shows the elements necessary to make a program for
desirable processing; we have only to make the data-flow graph by
combining these elements in the vacant space of the screen.

Several examples of elements are shown in Figure 74[58]. The
source (a) represents the input, the sink (b) the output, and the
primitive (c) the node in the data-flow graph. Also, the equation
block (d) represents not only the input and output data but also the
contents of the processing written in the central column. The branch

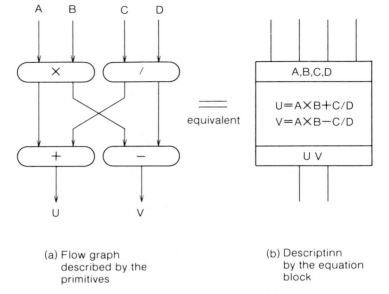

(a) Flow graph
    described by the
    primitives

(b) Descriptinn
    by the equation
    block

Figure 75. Representation of processing by primitive equivalent t_i data-flow graph.

block (f) shows the conditional branch. The common function (g) corresponds to the subroutines in the flow chart.

This language has several advantages:

(1) The parallel processing of the program can be understood.
(2) Two forms of program representation can be used; the program can be written in the form not only of the data-flow graph but also of the equation, as shown in Figure 75. The left-hand side shows the representation by the data-flow graph, and the right-hand side by the equation based on the equation block element (d), in which A, B, C and D are the inputs, U and V are the outputs. The equations written in the central column represent the necessary processing.
(3) The program can be written in a hierarchic representation and, accordingly, the outline of the processing can be understood at a glance. An example is shown in Figure 76.

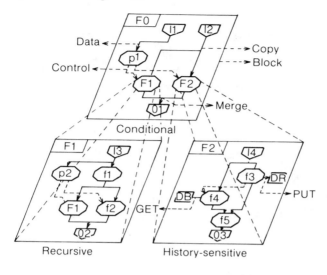

Figure 76. Hierarchic representation of processing in diagrammatical language

This language system has been proposed together with a compiler to convert the equation form representation into the dataflow graph also proposed.

5. Multiprocessor configuration and the performance of the prototype

The multiprocessor configuration is suitable, as mentioned earlier, for the data-driven processor to increase the calculation speed further. Figure 77 shows two examples of the multiprocessor configurations. Usually, a network connection is used as a unit for four processors.

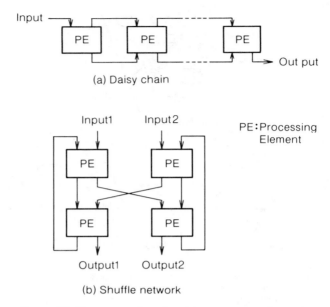

(a) Daisy chain

PE:Processing
Element

(b) Shuffle network

Figure 77. Two examples of multiprocessor configuration

A prototype of data-driven processor was constructed of several tens of breadboards. The packet transfer time between the neighbouring latches, shown in rectangles in Figure 69, was about 250 ns/packet. This means that the maximum processing rate is 4 MIPS if the processing is performed in every packet transfer time. Generally, the maximum processing rate is hardly realized.

Now, a test program is shown in Figure 78[57]. The normalized processing rate observed by executing the test program is plotted in Figure 79[57] against the generation number. The processing rate increases with generation number increase up to 4, though it is 0.1 for one generation. The processing rate levels off at 0.4 for generations above 4. This shows that the circular pipeline is not completely filled with the data packets of one generation. As mentioned above, it is obviously that the parallel processing in the circular pipeline is carried out completely at the maximum rate when every process is filled with a data packet. In the example of

Figure 78. Test program for prototpye data-driven machine

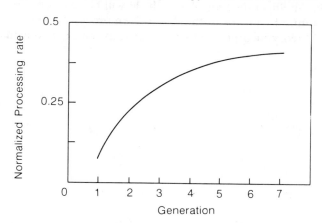

Figure 79. Example of performance of prototype data-driven machine

Figure 79, the pipeline is filled almost completely with the data packets of 4 generations. Thus, the processing rate levels off at 0.4 at higher generations above four. This value corresponds to 40% of the theoretical processing rate. This is quite a high level, which is difficult to attain. The processing rate would be increased much when an appropriate single chip processor is developed. Moreover, it can be expected that the performance of the data-flow architecture is increased further by adoption of the multiprocessor configuration. As a trial, adopting the shuffle-net configuration with four processing elements as shown in Figure 77, we obtained the processing rate four times as large as that of the single processing elements. Thus, the processing rate increases according to the number of the processing elements. This is the characteristic of the data-driven architecture.

6. Conclusion

A brief review was presented on the state of research on the data-driven architecture and its language system. The author believes that the intelligently controlled vehicle will be realized some time in the future by means of the data-driven processor and advanced control theory incorporated with improved sensors and actuators.

References

1. H. Harashima and T. Sasayama, J. of IECE (Institute of Electronics, Information, and Communication Engineers) **89**-5, 453 (1986), in Japanese.
2. Y. Hata, and S. Washino, J. of JSAE (Japanese Society of Automotive Engineers) **40**, 9, 1091 (1986), in Japanese.
3. H. Harashima and S. Washino, J. of SICE (Society of Instrument and Control Engineers), **25**, 11, 1023 (1986), in Japanese.
4. Proceeding of the symposium on control technology sponsored by JSAE (1983), in Japanese.
5. ibid. (1984).
6. ibid. (1986).
7. J. F. Cassidy, SAE Paper 770078 (1977).
8. J. F. Cassidy, M. Athans, and Wing-Hong Lee, IEEE Trans. **AC-25**, 5, 901 (1980).
9. A. R. Dohner, SAE Paper 780286, (1978).
10. A. R. Dohner, Automatica, **17**, 3, 441 (1981).
11. K. Matsumoto, T. Inoue, K. Nakanishi, S. Matsushita, S. Koganemaru, and H. Ooshika, SAE Paper 780590, (1978).
12. K. Ikeura, A. Hosaka, and T. Yano, SAE Paper 800056 (1980).
13. H. Harashima ed. Car Electronics Subsystem (Chunichisha Publishing Co. Ltd. 1987), Chap. 1, pp. 62–73, in Japanese.
14. T. Takahashi, T. Ueno, A. Yamamoto, and H. Sanbuchi, SAE Paper 850291 (1985).
15. T. Tabe, K. Matsui, T. Kakehi, and M. Ohba, IECON 85, 385 (1985).
16. T. Nakagami and Y. Asayama, Internal Combustion Engines, **25**, 317, 39 (1986), in Japanese.
17. T. Tabe, N. Ohka, H. Kuraoka, and M. Ohba, IECON **85**, 390 (1985).
18. H. Takahashi, Y. Eto, S. Takase, S. Murakami, and M. Maeda, Proceedings of SICE 87, 1241 (1987).
19. S. Soejima and S. Mase, SAE Paper 850378 (1985).
20. Y. Hata, K. Ikeura, T. Morita and T. Abo, 6th Proceeding of Intl. Conf. on Automotive Electronics, 82 (1987).

21. T. Abo, K. Sawamoto, and K. Matsushita, JSAE Technical Paper 861049 (1986), in Japanese.
22. T. Sasayama, S. Suzuki, M. Amano, N. Kurihara, S. Sakamoto and S. Suda, IECON 85, 68 (1985).
23. A booklet for 87 Tokyo Motor Show, Nissan Motor Co. Ltd., in Japanese.
24. M. Itoh, Automatic Control (Maruzen, Tokyo, 1981), pp. 148–152, in Japanese.
25. T. Fujii, Computer and Applications Mook (CORONA Publishing Co. Ltd., 1986), pp. 39–46, in Japanese.
26. S. Kuroiwa, T. Ohta, S. Kawai, System and Control, 26, 11, 716 (1982), in Japanese.
27. for example: M. Kobayashi, S. Takenouchi, Y. Kushiki and K. Sakamura, Proceedings of FJCC 87, 153 (1987).
28. for example: H. Terada, H. Nishikawa, K. Asada, S. Matsumoto, S. Miyata, S. Komori and K. Shima, ibid. 594 (1987).
29. T. Yuba, K. Hiraki, T. Shimada, S. Sekiguchi and K. Nishida, ibid. 578 (1987).
30. M. Amamiya, M. Takesue, R. Hasegawa, and H. Mikami, ibid. 602 (1987).
31. F. E. Coats and R. D. Freuchte, Intl. of Vehicle Design, Special Publication SP4, (1983), pp. 75–88.
32. Y. Nishimura and K. Ishii, SAE Paper 860415 (1986).
33. H. Andoh and M. Motomoti, SAE Paprer 870545 (1987).
34. H. Yamaguchi, S. Takizawa, H. Sanbuichi and K. Ikeura, SAE Paper 861389 (1986).
35. S. Washino, A. Takaoka and K. Ura, Trans. IECE, 57-B, 11, 689 (1974), in Japanese.
36. S. Washino, R. Nishiyama and S. Ohkubo, SAE Paper 860411 (1986).
37. S. Washino, H. Inoue, and R. Nishiyama, Internal Combustion Engines, 26, 334, 63 (1987), in Japanese.
38. S. Washino and R. Nishiyama, Trans. of JSAE, No. 33, 41 (1986), in Japanese.
39. H. Hasegawa, J. of JSAE, 37, 9, 986 (1983), in Japanese.
40. C. F. Aquino, SAE Publication, SP-487, pp. 1–16 (1981).
41. M. F. Bardon, V. K. Rao and D. P. Gardiner, SAE Paper 870569.
42. S. Washino and M. Tano, JSAE Technical Paper 861005 (1986), in Japanese.
43. S. Washino, S. Ohkubo and R. Nishiyama, JSAE Technical Paper 871050 (1987).

44. H. Iwano, T. Yamoto, and T. Ohta, JSAE Technical Paper 851003 (1985), in Japanese.
45. H. Andoh and M. Motomochi, Internal Combustion Engines, **25**, 317, 67 (1986), in Japanese.
46. T. Inoue and K. Aoki, J. of Automotive Engineers, **37**, 2, 141 (1983), in Japanese.
47. H. Hiroyasu, Proceedings of Japanese Society of Mechanical Engineers, No. 710-5 (1971), in Japanese.
48. N. Sawa, J. of Automotive Engineers, **27**, 4, 342 (1973), in Japanese.
49. S. Awano, Technology of Internal Combustion Engines, (Sankaidoo Publishing Co. Ltd., 1983), 4th ed., Chap. 3, pp. 80–92, in Japanese.
50. T. Nogi, S. Ohyama, T. Yamauchi and M. Fujieda, JSAE Technical Paper 852085 (1985), in Japanese.
51. H. Kubo, Text of Seminar on Fuel injection and its Optimization, Sponsored by Giken Information Center (1987), in Japanese.
52. R. Nishiyama, S. Ohkubo and S. Washino, JSAE Review, **8**, 2, 70 (1987).
53. R. Nishiyama, S. Ohkubo and S. Washino, Internal Combustion Engines, **26**, No. 334, 72 (1987), in Japanese.
54. T. Andoh, K. Iwagiri, M. Minoura, S. Itoh and T. Nagai, JSAE Technical Paper 842049 (1984), in Japanese.
55. H. Terada, H. Nishikawa, K. Asada, S. Matsumoto, S. Miyata, S. Komori, and K. Shima, Proceedings of FJCC, 594 (1987).
56. K. Shima, M. Meichi, S. Komori, T. Okamoto, S. Miyata, T. Tokura, M. Shimizu, H. Hara, H. Nishikawa, K. Asada and H. Terada, IECE Technical Research Report CAS86-137 (1986), in Japanese.
57. T. Tokura, Y. Tsuji, S. Takakura, Y. Nishikawa, H. Hara, M. Meichi, K. Komatsu, S. Yoshida, T. Okamoto, H. Nishikawa, K. Asada and H. Terada, ibid. CAS86-139 (1986), in Japanese.
58. H. Nishikawa, H. Terada, K. Komatsu, S. Yoshida, T. Okamoto, Y. Tsuji, S. Takakura, T. Tokura, Y. Nishikawa, S. Hara and M. Meichi, Proceedings of ICPP 87, 319 (1987).

Index

Japanese Technology Reviews